Radiation Effects on Electronic Systems

Radiation Effects on Electronic Systems

Henning Lind Olesen

Missile and Space Division
General Electric Company
Philadelphia, Pennsylvania

 SPRINGER SCIENCE+BUSINESS MEDIA, LLC 1966

Library of Congress Catalog Card Number 65-22183

© 1966 Henning Lind Olesen
Originally published by Plenum Press New York in 1966
Softcover reprint of the hardcover 1st edition 1966

ISBN 978-1-4899-5707-8   ISBN 978-1-4899-5705-4 (eBook)
DOI 10.1007/978-1-4899-5705-4

To my patient family, Birthe, Mette, and Robert

Preface

This book was originally prepared to serve as a reference book on radiation effects in electronics for electronics engineers at the General Electric Re-Entry Systems Department, who had been confronted with the task of designing space radiation hard electronic equipment. It served its purpose so well that I felt it could perhaps be of benefit to other electronics engineers in similar situations. Hence this volume was prepared. It consists of the original text with updating of those subjects which have been most influenced by the rapid expansion of activities in the radiation effects field throughout the country.

The book has been arranged in an order which seems most logical from a user's standpoint. The various kinds of nuclear radiation environments are described first. A terminology section is also included in this first chapter containing terms of which many are new and strange to the electronics engineer. Chapter 2 describes the atomic and nuclear interactions of radiation with matter and typical electronic devices. A large portion of this chapter is devoted to the effects on semiconductors, which are widely used by electronics engineers, but unfortunately very sensitive to nuclear radiation. The reader will soon discover that a thorough understanding and knowledge of these basic effect mechanisms will often suggest the design approaches which may be utilized to circumvent a particular radiation problem. Chapter 3 was prepared to make reference material available in the area of shielding. Shielding is often the first approach to overcoming the problems of radiation, and it is important to evaluate it as a design solution.

The reader will eventually be faced with the task of experimentally proving the design; Chapter 4 can serve as a planning guide.   Chapter 5 was included to show examples of actual applications of nuclear instrumentation. The reader may find examples which, with modification, can provide solutions to some current problem. Chapter 6, entitled "Electronic System Design Techniques," provides the reader with a suggested procedure for the design of radiation-hardened electronic systems. I would like to emphasize though that there is no one correct way to design radiation-hardened electronics. Much depends upon the engineer's creativity and ingenuity; there still is plenty of room for creative contributions in this radiation effects field which, after all, is only a few years old. Instead of the words "radiation-hardened" or "radiation-resistant," the reader may find the word "cosmitronic" useful. This term was formed from the words cosmic rays and electronic. The term is intended to be used as an adjective to signify that the particular electronic circuit or system has been constructed to withstand nuclear radiation. A cosmitronic amplifier, for instance, is an amplifier which has been designed to withstand nuclear radiation.

I am grateful for the permission of many technical journals and magazines to utilize information from papers written by outstanding experts in their respective fields. I particularly wish to thank the Nuclear Science Group of the Institute for Electrical and Electronic Engineers for permission to use significant portions of a paper published in IEEE Nuclear Science Transactions written by Dr. V. A. J. VanLint and E. G. Wikner of General Atomic, a division of General Dynamics Corporation. Dr. VanLint's many contributions on the subject of radiation effects on electronic parts have furthered the understanding of the problems and their solutions. His contributions are well known and recognized by all of us working in this field, and I am happy to be able to include this paper. The general field of radiation effects on electronics is so new that little information has as yet appeared in book form. The reader will find this reflected in the fact that the majority of the references are from technical journals.

Finally, I would like to extend my thanks to the many people at the General Electric Company's Re-Entry Systems Department who helped toward making the book a reality.

Special mention goes to Dr. R. T. Frost, Dr. H. Gerardo, E. R. Rathbun, W. Rowland, and D. Johnson for their help in editing the book. Thanks also are due to J. Bohuslaw, J. Barney, and M. Berkowitz for their encouragement during the preparation of the book; to D. Berstein, J. Duncan, and J. Silver for their contributions to Chapter 6; to J. Barney and D. Hendershott for their editorial and technical suggestions; and to Miss J. McGuire for her patience in typing the manuscript.

Henning Lind Olesen

Contents

CHAPTER 1

Radiation Environments   1

Description of the Environments   1
   Near and Solar Space   1
   Neighborhood of a Nuclear Space System   12
   Proximity of a Nuclear Burst   15

Radiation Terminology   16
   Some Radiation and Energy-Absorption Terms   17
   Particle Radiation   20
   Electromagnetic Radiation   23
   Method of Describing Gamma Exposure   24
   Comparison of Radiation Fields and Energy Absorbed   25

Analysis of Types of Radiation   27
   Electrons   27
   Gamma Rays   29
   Protons   30
   Neutrons   31

References   32

Bibliography   32

CHAPTER 2

Radiation Effects   35

Types of Radiation Effects   35
   Transient Ionization Radiation Effects   35

Displacement Radiation Effects.   38
Chemical Radiation Effects   42

Radiation Effects on Semiconductor Devices   43
Introduction.   43
Basic Processes   44
Semiconductor Operation.   45
Displacement Effects   54
Displacement   Effects   on   Various   Semiconductor
Devices   59
Ionization Effects.   76
Surface Effects   85

Other Radiation Effects   91
The Electromagnetic Pulse Effect.   91
Gamma and Neutron Heating.   92
Gamma-Induced Charge.   96

Summary of Material Effects   97
Radiation Effects   97
Special Circuit Design Considerations for the Nuclear
Environment.   98

Radiation Effects on Man   100
Radiation Exposure and Biological Effects   105
Comparative Risks from Radiation Sources.   108
Somatic Effects   111
Hereditary Effects   112
Radiation Protection Standards.   113

References   115

Bibliography   117

CHAPTER 3
Radiation Shielding   119

Introduction.   119

Passive Shielding   119
Neutron Bombardment.   119
Gamma Radiation.   120

Active Shielding 122
    Electrostatic Shielding Against Space Radiation 122
    Magnetic Shielding Against Space Radiation 123

Shielding Against Long-Term Radiation 124
    Radiation Levels Received Within a Typical Space-
    craft 127
    Conclusions 129

References 130

Bibliography 130

CHAPTER 4
Experimental Facilities 131

Radiation Experimentation 131
Experimental Considerations 136
    Experimental Design 137
    Test Monitoring 138
    Spectrum Determination 139
    Characteristics of Ground Test Reactor 139

Operating Descriptions of Various Pulsed Radiation
    Facilities 142
    KEWB (Liquid-Fueled Homogeneous Core-Reflected
    Pulsed Reactor) 142
    TRIGA (Solid-Fueled Heterogeneous Core-Reflected
    Pulsed Reactor) 144
    Bare Critical Assemblies 145

Facilities for Transient Ionization Effects Testing . . . 150
    Linear Accelerators 151
    Flash X-Ray Facilities 151

Dosimetry 152
    Neutrons 152
    Gamma and X-Rays 154

Reference 156

Bibliography 156

CHAPTER 5
The Nuclear Instrumentation System   157

   Introduction.   157

   Uses for Nuclear Instruments.   158
      Cosmology   158
      Nuclear and Particle Physics.   161

   Nuclear Instrumentation Designs.   161
      Spectrometers.   162
      Scintillation Detector   166
      High-Energy Gamma Ray Telescope   166
      Neutron Phoswich.   169
      Semi-Rad   172
      Material Analysis by Neutron Activation   172
      Airborne Scanner.   173
      Nuclear Instruments for Particle Physics.   175

   System Design — Missions in Space.   177
      Mariner II Instrumentation System   177
      Observatories in Space   181

   References   187

   Bibliography   188

CHAPTER 6
Electronic System Design Techniques   189

   Introduction.   189

   The Design Procedure.   190
      The Nuclear Radiation Environment.   190
      Shielding.   191
      The Electronic System   191
      The Circuit Analysis.   192
      Circuit Design.   193
      Simulation of the Environment   194

   Design Discussion.   195
      Electronic Parts Selection.   195

Active Devices.   196
Passive Devices.   199
Failure Analysis   201
Transient Analysis.   205
Packaging Design.   206

Design Examples   206
Digital Circuit Design.   207
Analog Circuit Design.   211
TIMM Record Amplifier   215
Pneumatic Controls and Logic   223
Timers.   223

References   223

Bibliography   224

Index. 225

CONTENTS

CHAPTER 1

Radiation Environments

DESCRIPTION OF THE ENVIRONMENTS

Electronic equipment which is sent into space may be exposed to several environments. In general terms, these environments fall into three classes, namely, (1) near and solar space, (2) neighborhood of a nuclear space system, and (3) proximity of a nuclear burst. Each of these environments will be discussed below with respect to its nature and characteristics. Typical radiation sources are given in Table 1-I.

Near and Solar Space

The penetrating nuclear radiation of near and solar space may be divided into cosmic radiation, trapped radiation (Van Allen belts), auroral radiation, and solar-flare radiation. Other, less important sources are the solar wind, which affects the magnetosphere, solar X-rays and neutrons. Electromagnetic radiation must also be considered.

Cosmic Radiation. Cosmic radiation consists principally of hydrogen nuclei (protons) whose velocity indicates that their kinetic energies are 1 to 10 BeV. It is impractical to shield against cosmic particles of such great energies.

The cosmic-particle flux is about 2 particles/cm^2-sec. This flux consists of 90% protons and 10% alpha particles (helium nuclei). The ionization dose rate attributable to cosmic radiation is about 4.5×10^{-4} rads/hr; that attributable to secondaries produced by the primary particles is about 10^{-3} rads/hr, the number being dependent on shield thickness. Most

*This section is taken from an article by the author which appeared in Electronics, Dec. 28, 1964.

TABLE 1-I

Typical Radiation Sources

Source	Radiation	Output
Natural radioactive materials		
Uranium	alpha	
Radium	beta	2.3 rads/hr
Radon	gamma	
Irradiated materials		
Cobalt-60	alpha	
Sodium	beta	to 10^7 rads/hr
Iodine	gamma	
Fission fragments		
Strontium	alpha	
Cesium	beta	10^9 rads/hr and up
	gamma	
From space		
Cosmic radiation	nuclei	2 particles/cm^2-sec
Solar flares	protons	1000 rads/hr
Van Allen belts	electrons	10 rads/hr
Accelerators	alpha	to 10^{11} rads/sec
	beta	
	gamma	10^{10} neutrons/cm^2-sec
	neutron	and up
Reactors	gamma	10^7 rads/sec
	neutron	about
		10^{15} neutrons/cm^2-sec

materials and electronic components are unaffected until an ionization dose of 10^4 rads and higher has been absorbed. Consequently, cosmic radiation does not pose a severe threat to the performance of electronic equipment.

Van Allen Belts.   Van Allen radiation consists of a great number of electrons and protons of various kinetic energies, all trapped by the earth's magnetic field.   The contour plot in Fig. 1-1 of the belts shows these in cross section. The particles form a toroid around the earth's geomagnetic equator. The charged particles — electrons and protons — are restricted to spiral paths surrounding magnetic field lines, and they continually drift between the southern and northern extremities of the magnetic field lines. A slight longitudinal drift is also evident; this makes the toroidal configuration complete.

The contour plot shows how the belt is represented as having an inner and outer belt. Flight data [1] show that the toroidal volume is permeated with protons and electrons. The inner belt flux peaks at about 2000 to 3600 km and the outer belt flux at 16,000 to 25,500 km altitude. It is postulated that the volume of low-particle flux separating the two peaks, the so-called "slot," is created by some phenomenon that reduces the lifetimes of the particles in this region.

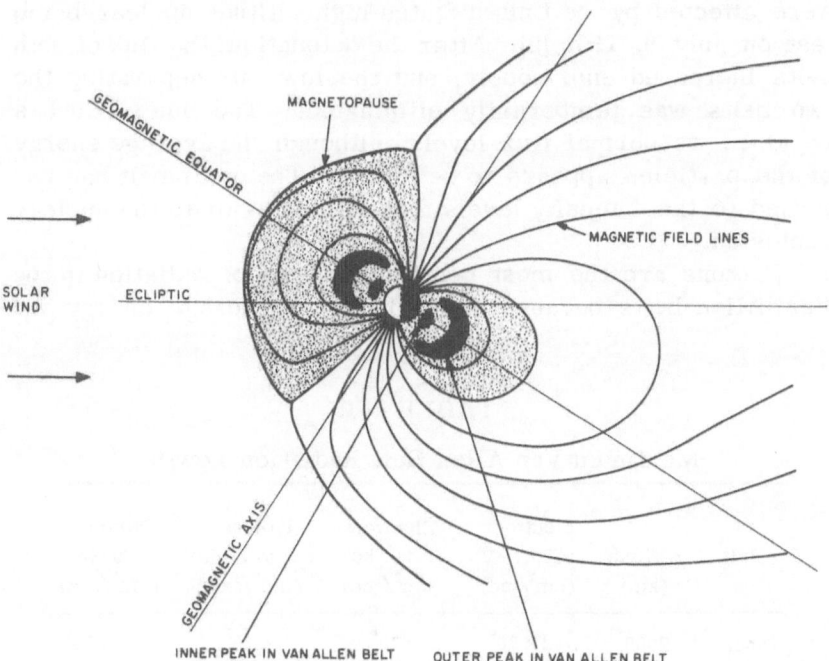

Fig. 1-1. Contour plot of magnetosphere showing Van Allen belts (after L. J. Cahill, Jr., Scientific American, March 1965, p. 59).

The inner Van Allen belt starts at an altitude of 400 to 1200 km, depending upon the longitude. It extends up to about 10,000 km, where it begins to overlap the outer belt. The inner belt extends from 45° north magnetic latitude to 45° south magnetic latitude. The region of highest flux occurs at 2000 km above the magnetic equator. The flux is strongly directional, with most particles moving perpendicular to the magnetic field. Table 1-III lists the fluxes at various particle energies. The energy gives an indication of the penetrating capability of the particular particle.

The outer Van Allen belt begins at an altitude of about 10,000 km near the magnetic equator and extends up to 60,000 to 82,000 km. The upper boundary is influenced by solar activity. The flux in the outer belt changes 10 to 100 times during the peak of a solar storm.

Normally the highest flux is found 16,000 to 25,500 km above the magnetic equator. Both the inner and outer belts were affected by the United States high-altitude nuclear-bomb test on July 9, 1962 [2]. After the detonation, the flux of both belts increased enormously, and the low flux separating the two belts was temporarily eliminated. The inner belt has regained its normal flux levels, although the average energy of the particles appears to be higher. The outer belt has returned to the intensity levels that existed prior to the nuclear explosion.

Protons are the most penetrating type of radiation in the Van Allen belts because of their higher average energy and

TABLE 1-II

Maximum Van Allen Belt Radiation Levels

Belt	Altitude (km)	Electrons > 20 keV (cm^2/sec)	Electrons > 200 keV (cm^2/sec)	Protons 0.1 to 5 MeV (cm^2/sec)	Protons > 60 MeV (cm^2/sec)
Inner	2000	$\sim 2 \times 10^9$	$< 10^8$	$< 10^6$	$\sim 4 \times 10^3$
Outer	25,500	$\sim 10^{11}$	$< 10^8$	$\sim 10^8$	$< 10^2$

TABLE 1-III

Estimate of Particle Flux in Van Allen Belt *

Particles	Energy	Intensity
Heart of inner zone:		
Electrons	> 20 keV	$2 \times 10^9/cm^2$-sec
Electrons	> 600 keV	$10^8/cm^2$-sec
Protons	> 40 MeV	$2 \times 10^4/cm^2$-sec
Protons	0.1 to 5 MeV	$10^6/cm^2$-sec
Heart of outer zone:		
Electrons	> 20 keV	$10^{11}/cm^2$-sec
Electrons	> 200 keV	$10^8/cm^2$-sec
Protons	> 40 MeV	$10^2/cm^2$-sec
Protons	0.1 to 5 MeV	$10^8/cm^2$-sec

*See [6] for detailed description.

their large mass. Shielding against protons is therefore difficult.

Although the average energy of electrons in the belts allows shielding, the secondary radiation they create is more troublesome. A medium-energy electron (0.5 MeV to 5 MeV) expends its energy mainly by ionizing materials; this results in a secondary dose of ionization. A high-energy electron (greater than 5 MeV) is ultimately absorbed by the bremsstrahlung process (braking or stopping electrons) which produces X-rays. X-radiation is more penetrating than the electrons and causes ionization radiation doses inside the vehicle. Maximum Van Allen belt radiation levels are given in Table 1-II.

Auroral Radiation. Auroral radiation, associated with the aurora borealis, is encountered between 65 and 70° north and south magnetic latitudes. The injection mechanism for these particles is not fully understood [3]. It affects space vehicles in polar orbits.

Auroral radiation is sporadic in nature, with the major flux delivered as electrons. Although an average yearly dose could run as high as 10^8 rads on the surface of a space vehicle, the secondary radiation amounts to only about 50 rads a year. Auroral protons have energies no higher than 650 keV, and a typical surface dose would be about 500 rads a year. Secondary radiation would be negligible.

Solar Flares. Solar storms produce large fluxes of energetic protons. These arrive at the earth within a day after a flare has been observed on the sun. The proton flux generally lasts less than 50 hr. If these are assumed to occur twice a year, about 10^9 protons/cm^2 would be received annually in the Van Allen belts.

The protons have a continuous energy spectrum. The ionizing dose caused by solar-flare radiation is approximately 10^2–10^3 rads a year for the interior of a space vehicle orbiting in the outer Van Allen radiation belt.

The Solar Wind. The spacecraft traversing the interplanetary space beyond the magnetosphere finds itself in a constant stream of charged particles (protons and electrons) emanating from the sun. This tenuous plasma stream was postulated as far back as 1896 by the Norwegian physicist O. K. Birkeland, but it was not until the advent of space flights that actual evidence of this solar wind was produced. The existence of this

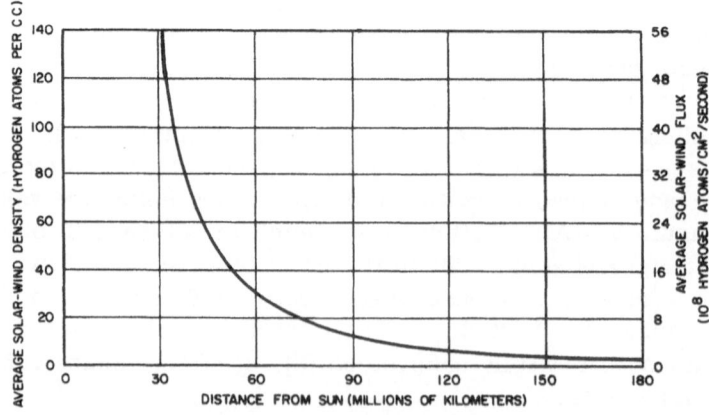

Fig. 1-2. Density and flux of solar wind (after E. N. Parker, Scientific American, April 1964, p. 72).

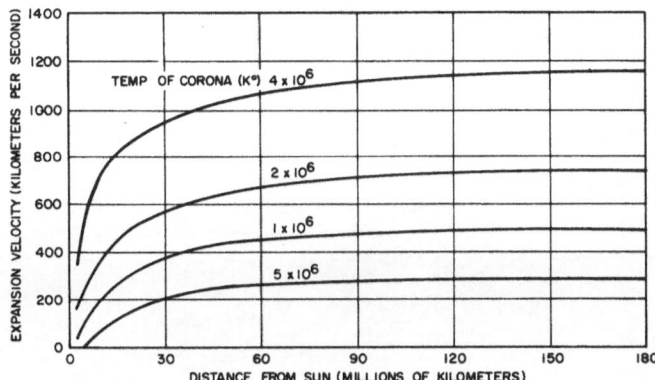

Fig. 1-3. Solar wind expansion rate (after E. N. Parker, Scientific American, April 1964, p. 72).

phenomenon was also surmised from the fact the gaseous tails of comets always point away from the sun, in response to the strong influence of the fast-moving solar wind. The solar wind is actually an extension of the sun's corona [4] and is responsible for the heating of the earth's upper atmosphere.

Figure 1-2 is a plot of density and flux of the solar wind. Figure 1-3 is a plot of the solar wind velocity, and depicts how the solar wind as it leaves the sun is rapidly accelerated, eventually attaining a constant velocity of from 200 to 1000 km/sec; the actual velocity depends on the temperature of the corona.

Various spacecraft, such as Lunik I, Lunik II, Explorer X, and Mariner II, have confirmed the fact that the solar wind streams continuously out from the sun radially in all directions. It is usually a steady wind, but becomes gusty during periods of high solar activity (solar flares). The velocity at earth's mean distance is approximately the expected 400 km/sec.

The influence of the solar wind is estimated to extend to at least 12 AU (astronomical units) from the sun, and much evidence indicates that it may extend as far as 20 to 50 AU. However, it is generally thought that it does not extend as far as 160 AU, or four times the distance to Pluto.

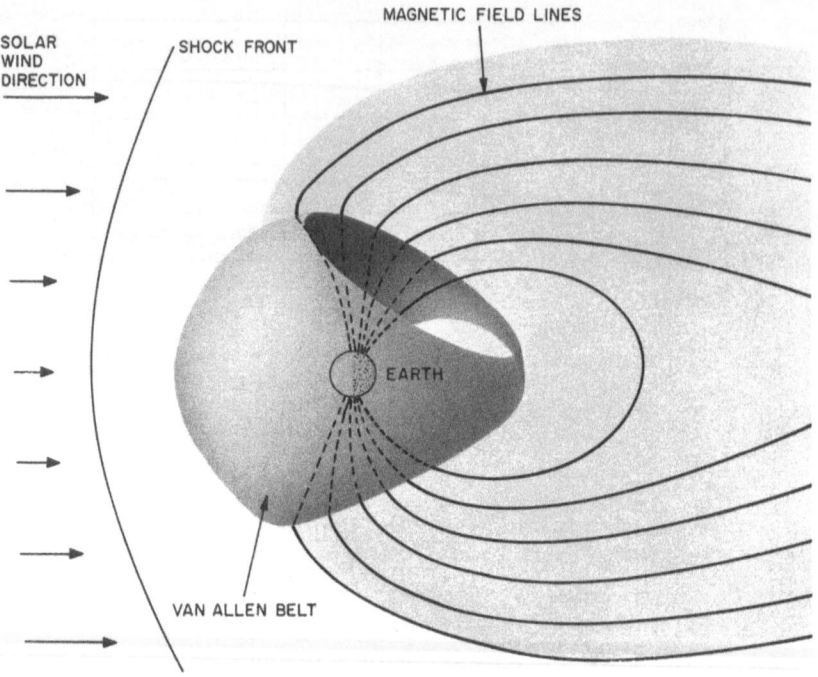

Fig. 1-4. Three-dimensional view of magnetosphere (after L. J. Cahill, Jr., Scientific American, March 1965, p. 66).

The Magnetosphere. Figure 1-1 shows the shape and the extent of the magnetosphere near earth. The inner magnetosphere is shown by the cross-hatched areas, and the inner and outer peaks of the trapped Van Allen belt particles by the black areas within the magnetosphere. Figure 1-4 is a three-dimensional presentation of Fig. 1-1.

As was the case with the solar wind, the actual understanding of how the earth's magnetic field is influenced by forces from the sun has been furthered extensively since the advent of space flight, which has enabled observers to send instrumentation into interplanetary space. The solar wind described above as consisting of charged particles influences the magnetosphere and causes the compression indicated by Fig. 1-1 on the sunlit side of the earth. The boundary of the distorted magnetosphere is termed the magnetopause. It is a comparatively thin (about 100 km) layer throughout which the interactions between the solar wind and the earth's geomagnetic field take place. Within the magnetopause the geomagnetic field drops abruptly and the direction of the field shifts by as much as 180° [3]. A current is actually thought to flow within the magnetopause; B. Sonnerup found evidence to support this theory [3] in his examination of Explorer XII magnetopause data.

Spacecraft such as Explorer VI, Explorer X, Explorer XII, Explorer XIV, Pioneer I, and Pioneer V all traversed the magnetopause and brought vivid evidence of an abrupt decrease in magnetic field strength as the magnetopause was crossed. Figure 1-5 shows the various spacecraft flights. Table 1-IV summarizes the particle fluxes in earth–sun space.

Electromagnetic Radiation. Other radiation in space that can affect materials are photons, i.e., electromagnetic radia-

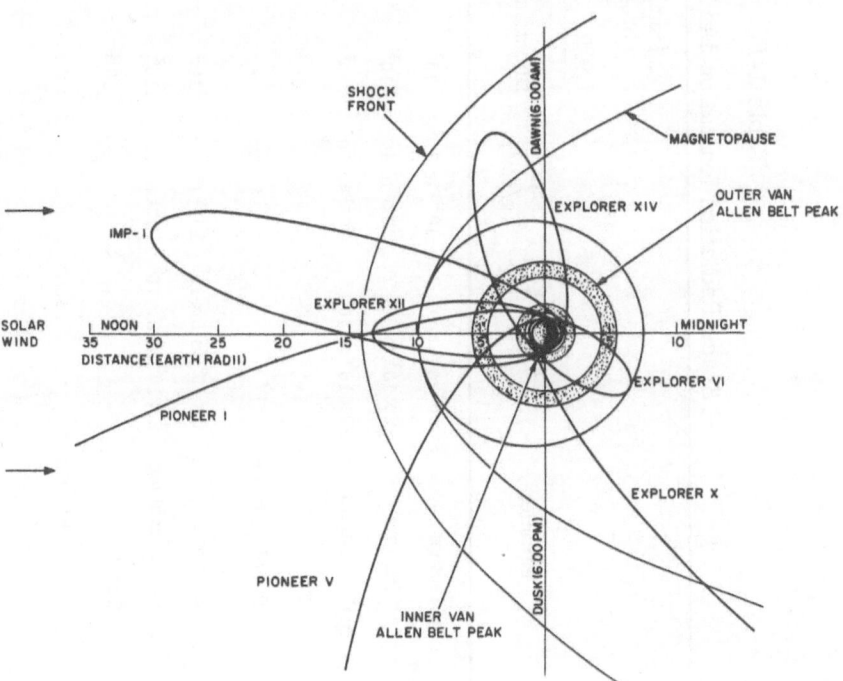

Fig. 1-5. Space flights surveying magnetosphere (after L. J. Cahill, Jr., Scientific American, March 1965, p. 60).

Table 1-IV
Space Radiation Ionization and Displacements *

Radiation source	Particles	Energy (eV)	Range (g/cm²)	Ionization (ergs/g-yr) at various exposures			Fractions of atoms displaced per year at various exposures		
				A Direct	B Through aluminum, 1 mg/cm²	C Through aluminum, 1 g/cm²	A Direct	B Through aluminum, 1 mg/cm²	C Through aluminum, 1 g/cm²
Inner Van Allen belt	(p) Protons	10^3 to 7×10^8	10^{-6} to 10^3	10^{12}	10^{11}	10^7	10^{-1}	10^{-5}	10^{-9}
	(e) Electrons	$<2 \times 10^4$ to 10^6	10^{-3} to 1	10^{14}	10^{14}	0	10^{-9}	10^{-9}	0
	(γ) Bremsstrahlung	$<2 \times 10^4$ to 10^6	10^{-1} to 10	10^7	10^7	10^7 to 10^8	$<10^{-13}$	$<10^{-13}$	$<10^{-13}$
	Total			10^{14}	10^{14}	10^8 max	10^{-1}	10^{-5}	10^{-9}
	Principal cause			e	e	γ,p	p	p	p
Outer Van Allen belt	Electrons	2×10^4 to 5×10^6	10^{-3} to 1	10^{14}	10^{14}	10^5	10^{-11}	10^{-11}	10^{-13}
	Bremsstrahlung	2×10^4 to 5×10^6	10^{-1} to 10	10^8	10^8	10^7	$<10^{-15}$	$<10^{-15}$	$<10^{-13}$
	Total			10^{14}	10^{14}	10^7	10^{-11}	10^{-11}	10^{-13}
	Principal cause			e	e	γ	e	e	e

Source	Component								
Solar flare (high-energy particles)	Protons	2×10^7 to 10^9	1 to 10^3	10^6	10^6	10^5	10^{-11}	10^{-11}	10^{-12}
	Electrons	$\sim 5 \times 10^4$	10^{-2}	10^8	10^8	0	0	0	0
	Bremsstrahlung	$\sim 5 \times 10^4$	1 to 10	10^3	10^3	10^3	0	0	0
	Total			10^8	10^8	10^5	10^{-11}	10^{-11}	10^{-11}
	Principal cause			e	e	P	P	P.	P
Solar flare (low-energy particles)	Protons	5×10^2 to 2×10^4	10^{-6} to 10^{-5}	0	0	0	<1	0	0
	Electrons	2×10^{-1} to 10	10^{-8}	0	0	0	0	0	0
	Bremsstrahlung	2×10^{-1} to 10	10^{-6} to 10^3	0	0	0	0	0	0
	Total			0	0	0	<1	0	0
Steady solar output	Protons	1 to 10^3	10^{-8} to 10^{-6}	0	0	0	<1	0	0
	Electrons	1	10^{-8}	0	0	0	0	0	0
	Bremsstrahlung	1	10^{-7}	0	0	0	0	0	0
	Total			0	0	0	<1	0	0
Cosmic rays	Protons	10^8 to 10^{19}	$>10^{-1}$	10^3	10^3	10^3	10^{-13}	10^{-13}	10^{-13}

*The ionization dose and displacement generation nominal values are accurate to within two orders of magnitude for typical electronic materials. This table is reproduced (with the permission of the ARS Journal) from L.D. Jaffe and J.B. Rittenhouse, ARS Journal, Volume 32, No. 3, March 1962, p. 331.

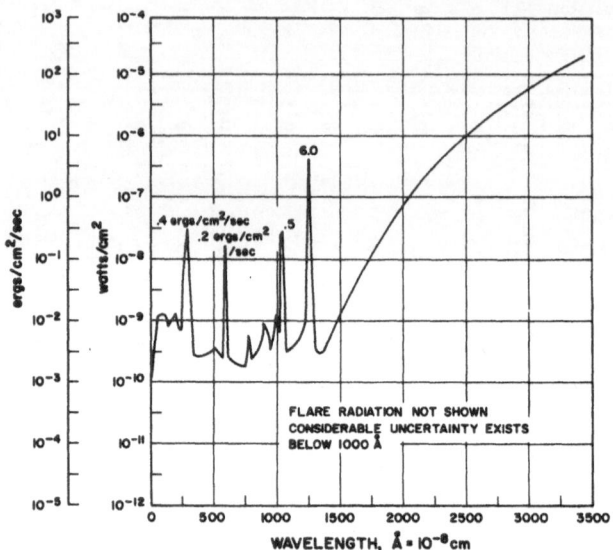

Fig. 1-6. Solar electromagnetic power at the earth mean distance (reproduced with the permission of L. D. Jaffe and J. B. Rittenhouse, ARS Journal, March 1962).

tion.   The sun is the primary source of electromagnetic radiation in the solar system. The intensity of this radiation varies inversely as the square of the distance from the sun. At earth's mean distance (1 AU), the total solar electromagnetic energy flux is 1.4×10^6 ergs/cm^2-sec or 0.14 W/cm^2 ± 2%. This is the number used in solar-cell power design. The distribution of the solar electromagnetic input just outside the earth's atmosphere is shown in Fig. 1-6.

Neighborhood of a Nuclear Space System

Radioisotope generators are being used to provide power for some space vehicles. These generators have low output per unit mass and are, therefore, restricted to applications requiring power levels of less than 1 kW. The electronics designer must be concerned with gamma radiation from these sources.

The flux level and primary gamma energy depend upon the type of isotope involved. Generally the gamma-dose rate for a radioisotope source is low enough so that the threshold for damage by ionizing radiation will not be exceeded in electronic equipment.

As power requirements in space vehicles increase beyond the kilowatt level, the radioisotope source becomes costly and

Table 1-V

Nuclear "Reactor-in-Flight" System (SNAP System)
Parameters

Characteristic	SNAP 10A	SNAP 2	SNAP 8
Power output (kWe)	0.5	3.0	30–60
Reactor power (kWt)	32	50	450–600
Efficiency (%)	1.96	6.0	—
Energy conversion	Ge—Si (thermo-electric)	Mercury Rankine	Mercury Rankine
Voltage	28.5 DC	120 AC	—
Boiling temperature (°F)	—	900-930	—
Average hot-junction temperature (°F)	900	—	—
Radiator temperature (°F)	615	600	—
Radiator area (ft²)	62.5	120	—
Unshielded weight (lb)	525	750	—
Radiation at base, gamma rays at 15 ft	10^7	10^7	10^7
Radiation at base, $(nv)t$, at 15 ft	10^{13}	10^{13}	10^{13}
Overall length (ft)	10	13	—
Diameter at base (ft)	5	5	—
Availability	1964	1966	1967

difficult to produce. It also becomes difficult and hazardous to transport the power plant. Nuclear power reactors become more attractive when power levels from 100 kW to many megawatts are required to operate for months or years. Nuclear reactors also can be used directly for propulsion. For direct propulsion, however, higher power levels are required—1000 MW and up. These are still far from ready to be used in flight. Table 1-V shows typical parameters.

Although the operation of nuclear reactors in these two applications differs somewhat, the emission of highly penetrating nuclear radiation is common to both. It is desirable to examine the radiation environment created by such a power plant.

To the electronic-equipment designer, two important types of penetrating radiation emanate from a nuclear reactor beyond the confines of the reactor core. One consists of ionizing gamma radiation of fluxes of about 10^4 rads (carbon) an hour. The other consists of neutron particles which, because they are uncharged, can penetrate materials; when these collide with the atoms of a material they change the material's mechanical and electrical characteristics, as described in Chapter 2.

The neutron environment is expressed as the total integrated dose per unit area. For a typical 100 kW nuclear reactor with a mission time of 10,000 hr, if some shielding is interposed between the reactor and the electronic system, the following is a typical radiation exposure for the electronic system: total neutrons with kinetic energies higher than 10 keV, 10^{13} neutrons per square centimeter; integrated ionizing dose 10^8 rads(C); ionizing dose rate, 10^4 rads(C)/hr (average gamma energy, 1 MeV).

Because electronic systems are usually located in the vicinity of the reactor cooling system, they operate in a high ambient temperature and the component temperature is increased further as the ionizing radiation expends its energy directly in the component materials. The nuclear reactor's design is a major determinant of its neutron- and gamma-leakage flux from a particular space vehicle's reactor and of the energy spectrum of the neutron flux. Fuel elements and coolants also influence the leakage flux.

The space vehicle's design can also affect the flux reaching the electronic system. Shadow shielding — reduction of radiation in one direction — reduces the flux in the direction of the electronic system, but the reduction of radiation by the inverse square of the distance can also be used to advantage.

Where a nuclear rocket is used directly for propulsion of the space vehicle, the comments about the nuclear power reactor still hold. However, the propulsion system is operated in different modes. When the velocity is to be changed, it is programmed for the required thrust, generating very high gamma and neutron fluxes during these short periods. This introduces transient ionization effects in the electrical behavior of the electronic system, caused by high ionization rates. These short-term electrical effects produce temporary degradation of performance. They can introduce false trigger signals in switching circuits, and result in a system malfunction.

Proximity of a Nuclear Burst [5]

A large portion of the nuclear explosion energy is received at a distance as thermal energy. For altitudes below 100,000 ft it amounts to about 35%, which is largely absorbed by the air surrounding the blast sphere. At altitudes above 100,000 ft the thermal energies approach 50% of the explosion energy and are not attenuated except by the $1/4\pi r^2$ factor.

The other large portion of the energy, about 50%, is contributed by the blast and shock associated with the explosion. Again these effects are most prevalent in the atmosphere and are nonexistent outside the atmosphere.

The remaining 15% is released as various nuclear radiations, and of these only 5% constitutes the initial nuclear radiation which for a moving space vehicle is generally the only radiation of interest. For a typical close-proximity explosion the vehicle will receive a large neutron dose and a high-intensity gamma pulse. The energy released per fission in a typical nuclear explosion is given in Table 1-VI. During the first microsecond of the explosion, 10^{22} such fissions take place. The equivalent of 1 kiloton of TNT amounts to a total of 1.45×10^{23} fissions.

Table 1-VI

Energy Released per Fission in a Typical Nuclear Explosion [5]

Type of energy	Energy (MeV)
Delayed:	
Kinetic energy of fission fragments	165 ± 5
Beta particles from fission products	7 ± 1
Neutrinos from fission products	10
Gamma rays from fission products	6 ± 1
Initial:	
Instantaneous gamma-ray energy	7 ± 1
Kinetic energy of fission neutrons	5 ± 0.5

For high-altitude explosions the center of the burst can be considered an isotropic source of nuclear radiation and the dose at the space vehicle can be calculated easily by the $1/4\pi r^2$ factor. Because the slower-velocity neutrons will arrive later than the initial ionizing gamma radiation, a neutron pulse stretching will also take place for the various energies or velocities of the neutrons. At distances greater than 90 miles from the center of a burst neutrons can be discounted as a source of radiation, since they have either been absorbed or scattered. The gamma radiation, on the other hand, is always able to induce transient ionization effects, and it is important to design the electronic circuits for tolerance against this effect.

Reference [5] should be consulted for a more comprehensive discussion of the Nuclear Burst.

Radiation Terminology*

Due to the acceleration of work in nuclear technology, and specifically in the area of radiation effects on materials and

*Taken in part from Memorandum No. 10, Radiation Effects Information Center, Battelle Memorial Institute, Columbus, Ohio, May 1959.

systems, a Radiation Effects Information Center (REIC) was
set up at Battelle Memorial Institute, Columbus, Ohio, to
serve as a central file for all radiation effects information
and to standardize the radiation effects terminology. The
institute is an excellent source of quantitative radiation effects
data, which the reader is urged to utilize.

There are two basic types of radiation that can disturb
the operation of electronic equipment: electromagnetic (speed
of light) radiation and particle radiation. The following are
some commonly used radiation terms. Under particle radiation, neutron terms predominate because neutrons in their uncharged state threaten great damage to electronic systems.
Neutrons can travel farther from the source, and penetrate
deeper into materials, than any other particle of comparable
energy.

Some Radiation and Energy-Absorption Terms*

Beta Rays: High-energy electrons ejected from a nucleus.

Protons: Hydrogen nuclei.

Alpha Particles: Helium nuclei.

Neutrons: Neutral particles with about the same mass as
protons.

Compton Electron: An electron which, after interacting with a
gamma photon by the Compton process, is removed from
its atom and moves through the material's crystal structure as a negative current carrier; in the case of electrons with kinetic energy > 200 keV, it moves as a displacement-producing particle.

Photon: As the wavelength of gamma and X-radiation decreases to below an arbitrary threshold of 0.124 Å or
0.124×10^{-8} cm, the behavior of electromagnetic radiation resembles that of particles. These particles are
called photons, each with an energy in ergs equal to
Planck's constant ($h = 6.625 \times 10^{-27}$ erg-sec) times the
frequency in sec^{-1}.

Compton Effect: The effect resulting from a collision between
a photon and an electron in the atomic structure of a
material. The electron may receive enough energy to

*Taken in part from an article by the author in Electronics, Dec. 28, 1964.

leave its atomic orbit and become a free electron, thereby ionizing the atom. The photon proceeds on a deviated
course and with a lower energy.

Gamma Rays: High-frequency electromagnetic radiation orgi-
nating from the nucleus as differentiated from X-rays,
which originate outside the nucleus. Energy levels are
expressed in MeV.

Million Electron Volts (MeV): A measurement of a photon's
energy, where 1 MeV corresponds to 1.6×10^{-6} ergs.
For most materials, the photon's absorption probability
can be assumed to be constant from about 500 keV to
about 2 MeV.

Cosmic Rays: Very high-energy particles which permeate
space (see Environments).

Van Allen Belts: Trapped particles found in two zones about
the earth's geomagnetic axis. The lower maximum
consists of protons and electrons, the upper maximum is
predominantly electrons.

Roentgen: Often abbreviated R or r, this term specifies the
amount of ionizing radiation (X-ray or gamma radiation)
released at standard temperature and air pressure. It
is the quantity of radiation that produces 2.083×10^9 ion
pairs/cc of air at standard pressure, 760 mm Hg, and
standard temperature, 25°C or 77°F at sea level. The rate
of energy release is expressed in roentgens per time unit.

Roentgen per Hour: The intensity of a field of ionizing radiation
which will deliver one roentgen per hour.

rad (m_0): The quantity of ionizing radiation that releases 100
ergs of energy per gram of the material. It is important
to specify the material when this term is used. Because
carbon is a common reference material, C is often found
in the parenthesis.

rem: Roentgen equivalent man. This term corresponds to
the quantity of radiation that produces biological damage
(see Radiation Effects on Man, Chapter 2).

Linear Absorption Coefficient (LAC) or μ: Each material has
a characteristic coefficient that expresses the efficiency
with which that material absorbs ionizing radiation. It
can be approximated as constant for most materials
between 0.5 MeV to 2 MeV photon energy. It appears in

the following equation: $I = I_0 \exp(-\mu x)$, where I is dose rate of dose per unit area after penetrating x cm of material, and I_0 is the initial dose rate of dose per unit area.

rep: A term seldom used with respect to electronic circuits. It is an acronym for roentgen equivalent physical, and expresses roentgens in terms of absorption of ionizing radiation by tissue. This absorption is equivalent to 93 ergs/g tissue.

barn: 10^{-24} cm^2, the unit for cross section, a measure of the probability that a nuclear interaction will occur.

ergs/g (of material m_0): The amount of energy absorbed in the form of ionizing radiation per gram of a particular material. Carbon or water is often used as a reference.

Rate of Energy Absorption: Ergs/g (m_0)-sec. The rate of energy absorption of material m_0.

Photon Flux: Flow of radiation. Assuming an average energy of 1 MeV, this flux is often expressed in terms of absorbed dose in carbon. It is expressed as MeV/cm^2-sec = 4.5×10^{-8} ergs/g (C)-sec.

Time-Integrated Dosage: MeV/cm^2, time-integrated dose of 1 MeV photons intercepting a unit area.

Absorbed-Dose Rate: The energy absorbed per unit time by a given material from the radiation field to which it is exposed.

Absorbed Dose: The time integration of absorbed-dose rate.

Exposure-Dose Rate: A measure of the radiation field to which a sample is subjected. It should be expressed in terms of the interaction of the field with a reference material (e.g., carbon), or in terms of flux.

Exposure Dose: The time integration of the exposure-dose rate.

Intensity: Synonymous with energy flux.

Threshold Dose: The minimum dose at which a change in some property of a material can be detected. Often the specific property involved is not given, and it is assumed that the most radiation-sensitive property was considered. This is also called the "threshold damage dose," although the property change is not necessarily deleterious.

25% Damage Dose: The dosage necessary to produce a 25% change in some property of a material.

Functional Threshold Dose: The minimum dose required to

change some property of a material or system to such
a degree that the system will no longer operate satisfactorily. Thus, the functional threshold for a given
material may vary with the application of the material.

Particle Equilibrium: This is said to exist in a region of a
material under irradiation when the energy absorbed per
gram is equal to the energy removed from the field of
radiation. In practice, this generally requires that the
region of interest be enclosed within a layer of effectively
the same material, having a thickness of at least the
maximum range of secondary particles generated by the
field. For gamma rays, this condition is usually referred
to as electronic equilibrium.

Particle Radiation

Neutrons are generally divided, often arbitrarily, into
energy ranges as follows:

Thermal: Neutrons that are in thermal equilibrium with their
environment, generally considered to be those whose
energy is less than approximately 0.1 keV. At room
temperature, the most probable energy is about 0.025 eV.

Epithermal: Generally taken to cover the range above thermal
energies.

Epicadmium: Neutrons whose energy is above the cadmium cutoff (see below), approximately 0.4 eV.

Slow: Neutrons in the energy range from 0 to 1000 eV.

Resonance: This is generally restricted to the range over
which resonance absorption in various materials can be
used for flux measurements. The range extends roughly
from about 1 to 10^4 eV.

Intermediate: Neutrons in the range of 10^3 to 5×10^5 eV.

Fast: Neutrons with energies greater than about 10 keV.

It is important to note that all of these neutron energy
ranges are rather loosely defined and there is considerable
overlap in some cases.

Kinetic energy is expressed in electron volts. It is a
measure of a particle's energy due to its mass and velocity.
Since the mass is known, it is useful to calculate the velocity
from the following equation:

$$V = \sqrt{c^2 - \frac{c^2}{\left(\frac{E_K}{m_0 c^2} + 1\right)^2}}$$

where $m_0 c^2$ is the particle rest energy (for neutrons, it equals 931 MeV), and E_K is the kinetic energy in electron volts.

The use of field-describing units has become common because of the difficulties in actually determining the energy absorbed in any given material. The problem of reporting and interpreting neutron-field measurements is much more difficult than is the case for a pure gamma field. Thermal neutrons are usually reported in terms of the nv_0 or $(nv_0) t$ (flux or time-integrated flux).

$\underline{nv_0}$: This unit describes a thermal neutron flux, preferably called a 2200 m/sec flux. These are neutrons with kinetic energies of 0.025 eV or, expressed in terms of their velocity, of 2200 m/sec. The associated unit of time-integrated flux is the $(nv_0) t$. The nv_0 is the neutron density normalized to the velocity v_0 of 2200 m/sec. When it is measured by foil techniques, a $1/v$ absorber and an absorption cross section corresponding to a neutron velocity of 2200 m/sec are used.

$\underline{\Phi}$: Symbol for the true flux. The true flux is the neutron density n multiplied by the average velocity $(\Phi = n\bar{v})$. The associated time-integrated true flux is Φt.

Fast neutron fields, i.e., neutrons with energies greater than 10 keV, shall be reported as the total neutron particles that cross each unit of area during exposure to a radiation source, neutrons $(> E)/\text{cm}^2$, where E is equal to some energy in keV or MeV, and the energy spectrum shall be given either in the form of a curve or identified with a known spectrum. The energy in keV is a measure of kinetic energy that the neutrons possess, and is therefore a measure of their penetrating power and ability to create displacement damage in materials. Dosimetry convention calls for the energy to be given as the lower limit of a spectrum. Other particles could replace neutrons in this term, for example, protons/cm^2, without changing the concept. Other neutron fields, i.e., neutrons less than 10 keV, shall be reported as neutrons $(< E)/\text{cm}^2$, where E is some energy less than 10 keV.

By convention, the velocity by which neutron densities are multiplied to obtain nv_0 for thermal neutrons is 2200 m/sec. However, except for the production of radioisotopes, thermal neutrons do not make a great contribution to the radiation-induced damage.   Most damage to the materials and components of interest to electronic engineers is caused by neutrons whose energy is above 10 keV.   Thus, for radiation-damage studies, dosimetry should be provided, which accurately measures the total dose of neutrons with energies greater than 10 keV. The most common way of measuring the fast flux is with a series of fission or resonance foils. The foils are made of materials that are activated by neutrons above a certain energy value.   Foils of various fissionable materials have been used for this purpose. With a series of foils having different threshold energies, one can describe the fast flux as a series of energy increments.   That is, the neutron energy distribution, and, hence, the field, is described in terms of neutron flux in several energy ranges.   It should be pointed out that, in calculating fluxes from foil activation, a good knowledge of the cross section of the foil for neutrons of a given energy is necessary.   Many of these cross sections are somewhat questionable at present, and so the cross-section values used should be stipulated.   Accuracies of the order of 13–50% are generally achievable.

Cadmium Cutoff: Cadmium can be used to absorb neutrons with energies below approximately 0.4 eV. At energies above this, the capture section for cadmium drops rapidly; below this energy the cross section is relatively high.

Cadmium Ratio: The ratio of the corrected activities in a bare foil to those in a cadmium-shielded foil. Thus, the ratio of the neutron flux below the cadmium cutoff to the flux above the cutoff is CR-1.

$1/v$ Detector: A foil-type detector whose cross section in the thermal region varies inversely with neutron velocity. The "$1/v$ flux" is then the neutron flux measured with such a detector. This flux will be the nv_0 flux, since the cross sections used are conventionally taken at 2200 m/sec.

When fast fluxes are reported in terms of the nv, it is implied that a specific and well-known average neutron velocity was used. Actually, the velocity range of fast neutrons is quite large, and so the nv is not applicable in practice to fast-neutron fluxes unless essentially monoenergetic neutron beams are being considered. It is recommended, therefore, that fast neutrons always be reported as integral flux in terms of neutrons/cm^2-sec or the corresponding time-integrated flux neutrons/cm^2 with a statement of the neutron energy distribution. It is desirable to show a curve illustrating neutrons/cm^2-sec-MeV vs. MeV, since knowledge of the neutron spectrum is important. The term (nv_0) should be reserved for thermal neutrons.

Electromagnetic Radiation

The following units of flux and intensity are sometimes used to describe exposures and are most useful in physically describing the radiation field:

Photon/cm^2-sec . . . . . . .Unit of gamma-ray flux

Photon/cm^2 . . . . .Time-integrated gamma-ray flux

MeV/cm^2-sec. . . . . .Gamma-ray intensity or energy flux

MeV/cm^2. . . . . .Time-integrated gamma-ray
                       intensity

For attenuated sources, it would be desirable to show a curve illustrating photons (or MeV)/cm^2-sec-MeV vs. MeV.

There are three commonly used units for describing the energy absorbed in a given material: erg/g; rad, defined as 100 ergs/g; and eV/g.

The unit ergs/g (C) is equivalent to the energy absorbed from an X- or gamma-ray field per unit mass of a limitingly small volume of carbon under conditions of electronic equilibrium. The graphite-wall CO_2-filled ion chamber is recommended as the working standard for measuring radiation fields in terms of this unit.

These units are easily interchangeable. The ergs/g and rad are commonly used by those working in the field of radiation effects. The radiation chemist generally prefers the eV/g, since his yields are usually reported in terms of G-values.(G-value is the number of molecules reacted per 100 eV absorbed.) Since these units describe energy absorbed in the sample, they are units of absorbed dose, rather than exposure dose. The absorbed dose is, of course, desirable in that it allows one to describe the damage to a material directly in terms of the energy absorbed in the material, regardless of the composition of the sample or the type of radiation field.

Method of Describing Gamma Exposure

It is generally desirable to report exposures received by all types of materials, components, and systems in a single system of units. Therefore, one should adhere to the ANP Advisory Committee for Nuclear Measurements and Standards' recommendation regarding gamma exposures, namely, that all gamma exposures be reported in terms of the field, the unit agreed upon being ergs per gram referenced to carbon [ergs/g (C)]. It should be reiterated that this is indirectly a measure of the field, and not a measure of the energy absorbed in a sample. In other words, the ergs/g (C) refers to the energy absorbed by the carbon-walled CO_2-filled ion chamber selected as the standard, and not the energy absorbed by the experimental sample.

In most work where gamma radiation is of prime importance, the material under examination is organic. Also, the energy range of the gamma field is generally such that the energy is deposited primarily by Compton interactions. For this reason, the relationship between the various units is simple and the conversions are easily made. The factors for converting the various units to ergs/g (C) are given below (equivalents for Compton contributions):

$$1 \text{ roentgen} \quad = 87.7 \text{ ergs/g (C)}$$
$$1 \text{ rep} \quad = 84.6 \text{ ergs/g (C)}$$
$$1 \text{ rad (tissue)} = 90.9 \text{ ergs/g (C)}$$
$$1 \text{ rad (water)} = 90.0 \text{ ergs/g (C)}$$

These conversions are particularly appropriate, since the dosimeters used will generally be either water solutions, tissue-equivalent materials, or CO_2-filled ion chambers.

It must be borne in mind that the carbon dose [ergs/g (C)] as it is used here is an exposure-dose unit and hence applicable to all materials. If one wishes to compare radiation effects on the basis of absorbed dose, it would be necessary to calculate the absorbed dose for each material from the description of the field.

In addition to roentgens, rads, etc., gamma fields are sometimes described by the photon flux or the energy flux. These fluxes may be readily converted to ergs/g (C) if the energy of the photons is known. In most cases of interest, the energy will be about 1 MeV. It is possible to relate number or energy flux to the exposure-dose rate as a function of photon energy. For example, assuming an average photon energy of 1 MeV, the following factors are obtained:

$$6.2 \times 10^3 \text{ photons/cm}^2\text{-sec} = 1 \text{ erg/g (C)-hr}$$
$$6.2 \times 10^3 \text{ MeV/cm}^2\text{-sec} = 1 \text{ erg/g (C)-hr}$$

Comparison of Radiation Fields and Energy Absorbed

Both of these methods of reporting radiation exposures, the description of the radiation field and the energy absorbed, have been and are being used by workers in the field. In engineering studies, the exposure in terms of the field is most commonly used. Perhaps the main reason for this is that in many cases the determination of the actual energy absorbed is practically impossible. The energy absorbed from a given field by a given material will depend upon the makeup of the field (gammas, neutrons, or mixed field) and the energy distribution of the field's components. Even if it were possible to report, unambiguously, the dose in all materials from all radiation fields, this would not be sufficient in many cases. Although the equal energy—equal damage concept may hold reasonably well for many organic compounds, it certainly does not hold for many other materials, such as semiconductors. In other words, although some materials can be damaged by any energy transferred from a radiation field, other materials are permanently damaged only when the energy-transfer process involves the nuclei of the atoms of the material, as in displacement processes. In this case, it is important to know how the energy is distributed among the neutrons making up the field. Therefore, any unit that gives only a statement of the energy absorbed will not be sufficient to convey all the information necessary to interpret damage data obtained in a neutron field. Such a unit would be suitable for pure gamma fields, since there is generally very little interaction between gamma photons and atomic nuclei. Furthermore, any unit that defines only the flux without stipulating the portion of the energy spectrum to which the flux refers will not be suitable. Tables VII and VIII are conversion charts.

TABLE 1-VII

Conversion Chart 1

To convert:	To:	Multiply by:
rad	ergs/g	100
eV/g	ergs/g(C)	1.6×10^{-12}
roentgen	ergs/g(C)	87.7
rep	ergs/g(C)	84.6
rad (tissue)	ergs/g(C)	90.9
rad (water)	ergs/g(C)	90.0
*MeV/cm^2	ergs/g(C)	4.5×10^{-8}
*photons/cm^2	ergs/g(C)	4.5×10^{-8}
*photons/cm^2	rep	5×10^{-10}
*rep/hr	neutrons/cm^2-sec	7.1×10^4
*rad/hr	neutrons/cm^2-sec	8.3×10^4
*rem/hr	neutrons/cm^2-sec	8.3×10^3
$n v_0$	rads/hr	4.2×10^{-6}
roentgen	rad (tissue)	1.036
rep	rad (tissue)	1.074
rem	rad (tissue)	91.290

*Assumed average energy of 1 MeV.

ANALYSIS OF TYPES OF RADIATION*

In this section we will analyze separately four specific types of radiation — electrons, gamma rays, protons, and neutrons — and discuss the types of radiation effects produced by each. The pertinent properties of these particles for radiation-effects analysis are summarized in Table 1-VIII for silicon and polyethylene matrices.

Electrons

Electrons of energy near 20 keV, such as are found in abundance in the Van Allen radiation belt, are incapable of producing atomic displacement in medium- and heavy-mass elements. For example, experiments have indicated that an energy of at least 13 eV is necessary to displace a silicon atom from its lattice position [7]. In a head-on collision, a 20-keV electron is capable of imparting only about one-tenth of this threshold energy to a silicon atom. On the other hand, electrons of energy near 200 keV are capable of providing this amount of energy to a silicon atom but not enough additional energy that the recoil atom itself can produce any secondary displacements. As a matter of fact, electrons in the energy region between 100 keV and 1 MeV are usually used to evaluate the threshold energy for atomic displacements. Higher-energy electrons, such as 40-MeV electrons produced by a linear accelerator, are capable not only of producing single displacements, but also of imparting to the recoil atom sufficient energy that it in turn can displace a large number of other atoms. Since the cross section for producing a displacement in an atom–atom collision is very large, it is possible that the secondary displaced atoms may be located fairly close together; hence, some of the manifestations of an irradiation by 40-MeV electrons may be different from those produced by 200-keV electrons. It is recognized, of course, that not many 40-MeV electrons are available in space. However, as is described in Chapter 4, such electrons are a convenient form of laboratory radiation and furthermore simulate quite accurately the effect of high-energy proton radiation.

*V.A.J. VanLint and E.G. Wikner, IEEE Transactions of Nuclear Science, NS-10, January 1963, with permission of IEEE.

TABLE 1-VIII

Pertinent Interaction Parameters

Bombarding particle	Silicon						Polyethylene		
	Energy loss (MeV-cm^2/g)	Range (g/cm^2)	Scattering length (g/cm^2)	Displacement cross section (barns)	Average number of displaced atoms per collision	Maximum number of displaced atoms per collision	Energy loss (MeV-cm^2/g)	Range (g/cm^2)	Scattering length (g/cm^2)
Electrons									
0.02 MeV	44	8×10^{-4}		0	0	0	50	7×10^{-4}	
0.2 MeV	7.7	4.5×10^{-2}		16	1	1.6	8.8	4×10^{-2}	
40 MeV	0.3	15		75	5	5000	3	15	
Gamma rays									
0.1 MeV			8						7
0.7 MeV			14						12
2 MeV			25						21
Protons									
10 MeV	35	0.16		3500	6	5×10^4	47	0.12	
100 MeV	5.9	10		350	7	5×10^5	7.3	7.2	
500 MeV	2.4	143		75	8	2.5×10^6	2.8	115	
Neutrons									
0.025 eV			20	0	0	0			0.3
1 MeV			10	4	2.5×10^3	5×10^3			2.1
14 MeV			25	2	3.5×10^4	7×10^4			9

All of these electrons dissipate more than 99% of their energy by ionization. Hence, they are very effective at producing transient radiation effects and chemical radiation effects. However, the displacement-producing effectiveness of these particles should not be ignored. For example, the number of displacements produced per unit path length by a 40-MeV electron is almost the same in silicon as the number of displacements produced by a 1-MeV fast neutron, in spite of the fact that a large fraction of the neutron energy is used for displacement production.

Gamma Rays

High-energy gamma rays interact primarily with the atomic electrons in matter. Depending upon the energy of the gamma rays, the predominant processes are photoelectric, Compton, or pair-production reactions. Gamma energies of 0.1 MeV are predominantly in the photoelectric regime. Gamma rays of 0.7 MeV, such as are typical of fission-product radiations, interact primarily via the Compton process. Gamma rays of 2 MeV are also in the Compton range, and these gamma rays are more typical of those emitted during the fission process. Higher-energy gamma rays, such as are produced by bremsstrahlung from high-energy electrons, interact primarily by positron-electron pair production. They are also capable of inducing photonuclear reactions. The nucleus recoiling from such a photonuclear reaction can frequently produce a large number of atomic displacements.

The secondary electrons which result from the photoelectric, Compton, or pair-production reaction can undergo processes similar to those discussed above under "Electrons." Hence, production of radiation effects by a gamma-ray beam depends primarily upon the efficiency with which those gamma rays are converted into secondary electrons and the subsequent interaction of those electrons with matter. In most practical cases the radiation effects of the photonuclear reactions can be ignored unless specific precautions are taken to enhance their importance by minimizing secondary electrons in the environment.

Protons

It is well known that high-energy protons are present in space, and a large fraction of the radiation damage observed in such devices as silicon solar cells must be ascribed to them. Protons of energies near 10 MeV are capable of producing a very-high-energy recoil atom. However, because of the long-range Coulomb interaction between the proton and the nucleus, most of the displacing collisions result in relatively small recoil energies, so that the average number of secondary displaced atoms is less than ten. The same type of process occurs with 100-MeV protons, with only a very slight enhancement of the average primary recoil energy, and hence of the average number of secondary displacements. The absolute cross section decreases significantly at the higher energy because of the form of the Rutherford scattering relationship. Protons of 500 MeV energy, in addition to displacing atoms by their Coulomb interaction, also have the capability of producing large nuclear interactions called stars. These stars may result in two types of interactions that can cause an enhanced radiation effect:

1. There are a large number of nuclear particles, including lower-energy neutrons and protons, which are emitted from a nuclear star. Since the protons, in particular, have significantly less energy than 500 MeV, their cross section for producing an atomic displacement is much higher than the cross section of the primary particles, and hence they may be expected to significantly enhance the displacement production. Detailed calculations indicate, however, that the flux of such secondary protons is too small to produce any large effect.

2. The residual nucleus, which results from the emission of high-energy and lower-energy particles in the nuclear star, recoils with a significant energy. This nucleus may have an energy of hundreds of kilovolts and hence is capable of producing a large number of displacements within a fairly small region of a solid lattice. Since the velocity of this nucleus is high enough that it may be partially ionized, a significant amount of the energy will be lost by electronic excitation and ionization of other atoms, hence producing transient and chemical radiation effects. Even so, the production of displaced

atoms from such a recoil nucleus is a significant contribution and can amount to as many, or more, total displacements than are produced by the primary proton beam via Coulomb interactions. As these displacements are produced by a single high-energy recoil atom, the radiation-induced effects so produced may be significantly different from those produced by a succession of lower-energy events. However, it would be surprising if most of the manifestations of these displacement clusters were not also observed in the high-energy electron or lower-energy proton irradiations, in which clusters of six or more displaced atoms are produced. In any case, as will be seen below, fast-neutron radiation produces recoil atoms with high energies and is thought to produce similar types of radiation effects.

Neutrons

Neutrons of <100 eV energy are obviously incapable of imparting a significant amount of energy to a lattice atom by inelastic collision. However, the absorption cross sections for thermal neutrons are significant and in many cases atoms can recoil from their lattice positions as a result of the emission of gamma rays associated with thermal-neutron capture. There are also some large-cross-section reactions, such as B^{10} (n, α), and Li^6 (n, α), which are exoergic and produce recoil atoms having large ionization and displacement-producing power. In the presence of fast neutrons and gamma rays, the effects of thermal neutrons can usually be ruled out, unless such high-cross-section material as boron, lithium, or a fissionable material is present.

Neutrons of >10 keV energy, such as are characteristic of the fast neutrons in a nuclear reactor, produce significant displacement radiation effects by elastic collisions with lattice atoms. These lattice atoms receive energies of the order of tens of kilovolts from such collisions and produce secondary displacement cascades amounting to hundreds of displaced atoms. In a typical reactor environment the displacement radiation effects are produced predominantly by the fast neutrons.

Neutrons of >14 MeV energy, in addition to producing displacement effects by the elastic scattering interaction discussed above, also impart to many atoms sufficient energy that the recoil may be partially ionized. Hence, in analogy with the recoils from nuclear star formation discussed above, a portion of the recoil energy is dissipated as ionization and electronic excitation and is manifested as transient or chemical radiation effects.

REFERENCES

1. B. J. O'Brien, "Radiation Belts," Scientific American (May 1963).
2. W. N. Hess, "Earth's Radiation Environment," Space/Aeronautics (November 1964).
3. L. J. Cahill, Jr., "The Magnetosphere," Scientific American (March 1965).
4. E. N. Parker, "The Solar Wind," Scientific American (April 1964).
5. S. Glasstone, "The Effects of Nuclear Weapons," U.S. Government Printing Office, U.S. AEC, Revised Edition, Feb. 64, pp. 1 to 27.
6. M. B. Baker, "Geomagnetically Trapped Radiation," AIAA J. 3(9):1569-1579 (Sept. 1965).
7. J. J. Loferski and P. Rappaport, Phys. Rev. 111:432 (1958).

BIBLIOGRAPHY

Space Radiation

Berkner, L. V., and Odishaw, H., Science in Space, McGraw-Hill Book Company (New York).

Hines, C. O., "The Magnetopause," Science 141 (July 1963).

Jaffe, L. D., and J. B. Rittenhouse, "Materials in Space Environment," ARS Journal (March 1962).

Johnson, F. S. (ed.), Satellite Environment Handbook, Stanford University Press (Stanford, California), 1961.

LeGalley, D. P., and A. Rosen (eds.), Space Physics, John Wiley & Sons, Inc. (New York).

McIlwain, C. E., "The Radiation Belts, Natural and Artificial," Science, Vol. 142, No. 3590, p. 355, Oct. 18, 1963.

Odishaw, H. (ed.), Research in Geophysics, The M.I.T. Press (Cambridge, Massachusetts).

Parker, E. N., Interplanetary Dynamical Processes, Interscience Publishers, Inc. (New York), 1963.

Rogers, S. C., "Radiation Damage to Satellite Electronic System," IEEE NS-10 (January 1963).

Van Allen, J. A. (ed.), "Radiation Belts Around the Earth," Scientific American (March 1959).

Collected papers on the artificial radiation belt from the July 9, 1962 nuclear detonation, J. Geophys. Res. 68 (February 1963).

Final Report on the Relay I Program NASA Report NASA SP-76 (1965), Chapter 6, p. 403 and Chapter 7, p. 429.

Memorandum No. 10, Radiation Effects Information Center, Battelle Memorial Institute (May 1959).

Nuclear Propulsion

Anderson, G. Montgomery, "Nuclear Reactor Systems," Astronautics & Aerospace
    Engineering (May 1963).

Johnson, P.G., "Beyond Apollo with Nuclear Propulsion," Astronautics & Aero-
    nautics (December 1964).

Jordan, W., et al., "Nuclear Rocket Stages Increase Saturn's Payload Capability,"
    Aerospace Eng. (May 1961).

Osmun, W.G., "Nuclear Interplanetary Mission Study," Lockheed Report NSP 64-34
    (March 1964).

Osmun, W.G., "Space Nuclear Power: SNAP 50/SPUR," Space/Aeronautics (Decem-
    ber 1964).

Pratt, P., "Aircraft Propulsion Systems in Evolution," Astronautics & Aeronautics
    (March 1965).

Nuclear Weapons

Glasstone, S., "The Effects of Nuclear Weapons," U.S. Government Printing Office.

CHAPTER 2

Radiation Effects

Radiation effects are produced when radiation energy is expended in a material. Radiation energy may be either of two kinds: electromagnetic or particles. Electromagnetic radiation is characterized by a velocity equal to that of light and has energy that can be determined by Planck's law (energy = hf), where h (Planck's constant) = 6.625×10^{-27} erg-sec and $f = c/\lambda$ = gamma-ray frequency in sec^{-1}. Gamma radiation, because of its extremely high frequency, is one of the most penetrating forms of electromagnetic radiation. The second basic form of radiation energy includes all particles moving at a velocity less than that of light. The particles may be charged, as protons and electrons, or they may be neutral, as neutrons.

TYPES OF RADIATION EFFECTS*

We will consider three general types of radiation effects: transient, displacement, and chemical radiation effects. Transient effects are due to electrons which have been excited or ionized by the radiation. Displacement effects are due to defects created by displacing atoms in a solid lattice. Chemical radiation effects are due to molecular rearrangements resulting from ionization energy deposition. Figure 2-1 relates the radiation to produced effects.

Transient Ionization Radiation Effects

Transient radiation effects are those manifestations of the interaction of radiation with matter which are associated with

*V.A.J. VanLint and E.G. Wikner [1], with permission from IEEE.

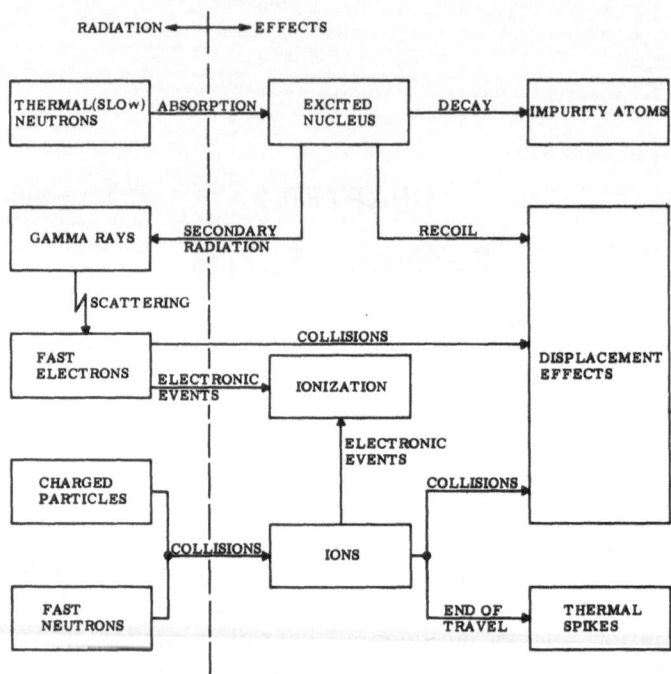

Fig. 2-1. Radiation effects (after V. A. J. VanLint and E. G. Wikner, IEEE, Transactions of Nuclear Science NS-10, Jan., 1963).

the excitation (including ionization) and de-excitation of electrons. Since transient effects are associated with changes in
electron states, they usually produce significant changes only
in the electrical and optical properties of materials, and in
most cases the perturbations are short-lived. Relaxation
times for electronic changes in most materials are usually
very short, and hence in many cases the basic transient effects
are a function (of power 0.5–1) primarily of radiation dose
rate, and they disappear soon after the irradiation ceases; hence
the term transient effects. It should be noted, however, that
some transient effects are long-lived (e.g., F centers produced
by populating an existing negative-ion vacancy in alkali halides
with an electron).

Transient effects have become an important problem in
connection with electronic circuits which must operate in an
intense nuclear radiation environment, particularly in military

electronics, which potentially may function near a nuclear detonation. Transient effects can be divided into three stages: (1) the excitation of electrons; (2) the perturbations produced in materials by the excited electrons; and (3) the electron deexcitation reactions. The perturbations and recombination reactions are frequently similar to those produced by optical excitation of electrons, e.g., photoconductivity.

Electron Excitation. High-energy, charged nuclear particles (e.g., electrons, protons, alpha particles, and charged fission fragments) produce electron excitation directly by interacting via the electrostatic Coulomb field with atomic electrons. Neutral particles may produce charged secondaries; for example, gamma rays deposit most of their energy as secondary electrons produced by the pair-production, Compton, or photoelectric process, and neutrons may produce ionizing recoil particles, especially in materials containing hydrogen atoms. The more energetic ionized electrons produce further excitation. Eventually, all of the ionized electrons, except the secondary electrons which may escape from a sample of finite extent, become slowed down to the point at which they are incapable of producing further excitation. They then may lose energy by other inelastic and elastic scatterings until they attain thermal energy or are captured.

It has been established that the electron—ion pair production in gases can be characterized in each gas by a unique efficiency having a value between 25 and 45 eV expended per pair produced, independent of the nature or speed of the primary ionizing particle. A similar efficiency has been measured in semiconductors, amounting to 3 to 4 eV expended per electron—hole pair generated. A reasonable extension of these data to other materials indicates that an amount of energy equal to two to four times the ionization potential of the material would be expended per ionization produced. In applying this rule, one must not include energy dissipated by neutral particles in nonionizing processes, such as neutron-produced recoil-atom motion below the energy at which atoms ionize.

Effects of Excited Electrons. The types of perturbations produced in materials by excited electrons (most of which are ionized) include the following:

1. In semiconductors, the densities of majority and minority carriers are changed; this results in conductivity changes, a decrease in reverse impedance of rectifying junctions, and generation of photovoltages at junctions.
2. In insulators, secondary electrons are emitted from surfaces; this produces net charges on conducting elements and in addition induces an internal space-charge distribution.
3. In gases, free electrons and ions are produced; these conduct electric current and frequently exhibit temporary net space-charge distributions.
4. The production of free electrons and holes in insulators is similar to that in gases, although the mobility of the free carriers is greatly reduced in the solids. Space-charge effects can also be much longer-lived.
5. Transparent materials may exhibit appreciable optical effects, such as change in transmission at certain wavelengths and emission of fluorescent radiation.

Electron De-Excitation. The electronic perturbations eventually are removed by various de-excitation processes. In semiconductors, electron–hole recombination takes place, assisted by defect centers, in 10^{-8} to 10^{-3} sec (10^{-6} sec in typical semiconductor parts). In conductors, the secondary emission current is replenished promptly by the attached circuit. In gases, the electrons may be attached to electronegative molecules (e.g., O_2), forming a negative ion, or they may recombine with a positive ion, or they may diffuse to a surface. The ions continue to affect the conductivity until they recombine, although they are less effective than free electrons. In insulators, the carriers may recombine rapidly (within $\sim 10^{-12}$ to 10^{-10} sec) or one or both type of carriers may become trapped at defect sites. Such trapping may be long-lived, as in the formation of F centers in alkali halides; F centers can be observed as a visible coloration. Space-charge distributions may also exhibit long relaxation time (as long as a number of days).

Displacement Radiation Effects

Displacement radiation effects usually involve the following occurrences:

1. A fast nuclear particle entering a material makes a close collision with the nucleus of an atom, imparting to it sufficient energy to displace the atom from its lattice site.
2. The displaced atom moves through the solid, losing energy in collisions with other atoms and displacing some of them. All of the displaced atoms eventually come to rest.
3. The lattice defects so formed may be thermally unstable, even at room temperature, and some of them may anneal, either annihilating the defect entirely or forming a secondary defect.
4. The defects influence various macroscopic physical properties of the material.

If the primary radiation is charged, as is the case, for example, with high-energy electrons, protons, and deuterons, the interaction between it and an atom to produce a displacement is usually via the Coulomb electrostatic force between the nuclear charge and the moving particle. Since for fast particles the impact parameter must be much less than the average radius of the K electron orbit in order to produce a displacement, it is inferred that the shielding effect of the orbital electrons is unimportant. A neutron produces displacements by interacting with the nuclear force field and hence interacts only at much shorter impact parameters than do charged primaries.

For the most part, a gamma ray produces displaced atoms by first generating, by the photoelectric, Compton, or pair-production process, secondary electrons which then produce the atomic displacements. High-energy (> 10 MeV) gamma rays can also produce some displaced atoms as recoils from photonuclear reactions; in beryllium or deuterium, gamma rays of lower energy ($\gtrsim$ 2 MeV) may be effective.

The cross sections for the pertinent interactions are fairly well known. The Coulomb interactions between charged particles and nuclei result in a spectrum of displaced-atom energies E varying as $1/E^2$. The limiting value is determined by the energy transfer in a head-on collision, which depends upon the energy of the primary particle and the mass ratio. The spectrum of energies of the displaced atoms is sharply peaked for low-energy atoms, even though the maximum possible energy transfer may be quite large.

For the case of a neutron producing an atomic displacement, the interaction is due to the short-ranged nuclear force field. This interaction can be approximated as a hard-sphere collision with a total cross section of the order of the elastic scattering cross section ($\sim$ 5 barns) for fast neutrons. The energy spectrum of the atoms which have been hit in this case is approximately uniform between zero and the maximum possible energy transferred; hence, the average energy imparted to the displaced atoms is approximately half of the maximum possible energy transfer, in contrast to the charged-particle-induced collision.

It is usually assumed that an atom can be displaced from its lattice site if it receives more energy than a certain threshold value, E_d ($\sim$ 25 eV in many materials). Most calculational models assume that the atom will not be displaced if it receives less than the threshold energy and will always be displaced if it receives more than the threshold energy. This assumption is obviously an idealization. In interpreting experiments on threshold effects, it must be known whether the experiment is sensitive to the lowest energy at which an atom can be displaced, even in a fortuitous event, or whether it is sensitive to an average energy over the displacement probability curve.

An atom that has received significantly more energy than is necessary to displace it will usually encounter other atoms in its motion through the lattice and perhaps will displace some of them. The interaction potential for this process has been studied extensively, and a reasonably good formulation for the problem can now be made. This interaction is a relatively long-range one, so that atoms receiving energies of tens of thousands of electron volts will travel distances of at most some hundreds of angstroms before coming to rest. They dissipate their energy by displacing other lattice atoms and inducing lattice thermal vibrations. A "rule of thumb" used to predict the average number of displaced atoms resulting from a primary collision is that one atom is displaced for each amount of energy of twice the average threshold energy dissipated as kinetic energy of recoil atoms.

It is well known that even at temperatures significantly below room temperature some radiation-induced defects are unstable. For example, in copper, annealing peaks, as manifested by changes in residual resistivity, have been observed at 17°K and possibly at lower temperatures. Some of the lowest-temperature annealing effects have been interpreted as manifestations of recombinations between interstitial atoms and vacancies which were originally close enough together to be within each other's strain fields. For the noble metals, the experimental data have been interpreted to show that interstitials move even in an unperturbed lattice at temperatures well below that of liquid nitrogen. Information available on silicon indicates that the vacancies move at temperatures between that of liquid nitrogen and room temperature. The available experimental information is sufficient to permit the conclusion that rarely, if ever, are primary radiation-induced defects observed at room temperature. It is also very likely, as in the case of silicon, that many secondary defects are present which are the result of association between primary-radiation-induced defects and impurities present before irradiation. Since there is here a complicated two-step process in which one participant is an impurity whose identity is in most cases unknown and whose presence may be unsuspected, care must be taken in drawing general conclusions about the manifestations of displacement radiation effects.

The physical manifestations of displacement-induced defects are many, and include the following:

1. Increase in the electrical resistivity of metals, particularly at low temperatures, as the result of the enhanced concentration of electron scattering centers.
2. Changes in the minority carrier lifetime, carrier mobility, and effective doping of semiconductors, as the result of the defect states introduced in the forbidden energy gap.
3. Changes in the mechanical properties of materials, as a result of the effect of the radiation-induced defects on the lattice.

4. Changes in the mechanical properties of alloys, result-
    ing from localized recrystallization and rearrange-
    ment collisions.
5. Changes in the thermal conductivity of materials, as a
    result of lattice defects which act as phonon and elec-
    tron scattering centers.

These and other manifestations depend upon the nature and
abundance of the defects, but usually a variety of interaction
mechanisms are involved. For example, the presence of a
defect in a lattice usually changes the lattice parameters and
mechanical properties. The defect will change the effective
carrier concentration in a semiconductor only if it actually
accepts or donates an electron, i.e., only if the Fermi level
is in the proper relation to the energy level of the defect. More-
over, the scattering of current carriers by the defect depends
on whether the defect is or is not charged. Hence, one must
make use of detailed information about the nature of the defect
and its charge state before being able to predict accurately the
perturbation that it will introduce into the material.

Chemical Radiation Effects

In certain systems, including gases, liquids, and organic
solids, the primary radiation effect is a chemical change. These
changes are usually brought about by the following steps: (1)
Radiation interacts with the atomic electrons, producing free
electrons and positive ions; (2) the positive ions undergo
secondary reactions, including ion clustering, charge exchange,
and ion exchange; (3) the free electrons lose energy by inelastic and elastic scattering, and may either recombine with
positive ions or attach to neutral or already negative molecules
to form negative ions; (4) negative and positive ions recombine,
and energy is released as kinetic energy of the resultant molecules; and (5) some of the recombination products may be
chemically active free radicals which will participate in
secondary chemical reactions.

As a result of these processes, the effect of radiation is to
change molecular configurations. Some of the reactions which
occur in air when it is subjected to ionizing radiation afford
simple examples; the primary ionization process is the forma-

tion of free electrons and positive ions, including N^+, N_2^+, O^+, and O_2^+. Experimental observations have shown that the N^+ and N_2^+ rapidly become associated with nitrogen molecules to form N_3^+ and N_4^+; the N_2^+ also rapidly exchanges charge with O_2 to form O_2^+; the N^+ reacts with oxygen to form NO^+. The free electrons usually attach to oxygen molecules forming O_2^-. Subsequent reactions between nitrogen and oxygen ions result in a variety of nitrogen oxides, including NO_2 and NO. As a result of a prolonged irradiation of a sample of air, there is an appreciable concentration of nitrogen oxides formed. This effect is a long-lived perturbation produced by ionizing radiation and is part of the subject of radiation chemistry, although the primary processes might be classed as transient radiation effects.

There are similar occurrences in liquids and organic solids. For example, irradiation of water produces hydrogen peroxide and free hydrogen. Polyethylene undergoes reactions in which hydrogen is released and bonds are made between adjacent polymer chains (cross links). Teflon undergoes a process in which fluorine and other decomposition products are evolved and the polymer chains are broken. The chain-breaking process in Teflon partly explains its rapid degradation in a radiation environment, as compared with the slight improvement in mechanical properties in polyethylene due to cross linking during irradiation. It has also been noticed that Teflon degrades less rapidly when irradiated in vacuum than when irradiated in air; it appears possible that one of the constituents of air (probably oxygen) contributes to the degradation.

Organic liquids undergo similar reactions, although frequently it is difficult to trace in detail what is happening because of the complexity of the liquids. However, the primary interactions of the radiation are known, and useful predictions can be made from this knowledge.

RADIATION EFFECTS ON SEMICONDUCTOR DEVICES

Introduction

As the need for design of radiation-resistant electronic equipment arose it immediately became apparent that the retention of solid-state electronics would be difficult because semiconductors are highly susceptible to radiation damage. This was discovered as early as 1948. Paradoxically the vacuum tubes which the semiconductors have replaced are much less sensitive to radiation damage. However, the desire to maximize scientific payloads aboard spacecrafts has dictated the requirements of smaller power consumption and smaller volume, which solid-state equipment provides. Although the lift-carrying capability of today's boosters has improved, it is still desirable to maintain a high performance-to-weight ratio for their payloads.

Thus a major portion of the activities in the radiation-effects field has been spent in the effort to improve the semiconductor performance during and after the influence of nuclear radiation. These activities have been broadly concerned with two endeavors: (1) research into the causes of radiation effects on semiconductors, in order to discover ways for minimizing the effects, and (2) search for better semiconductor devices insofar as performance when exposed to nuclear radiation is concerned.

The following will summarize these activities and describe the interactions and their effects on semiconductors. The section is important, since the reader will often find a thorough understanding of the effects physics of direct benefit in his design endeavors.

Basic Processes

The nuclear radiation will produce in the semiconductor materials two of the effects described in the section on types of radiation effects, displacement effects and transient or ionization effects. The third effect is the so-called Telstar effect, or surface effect, which is caused by radiation-induced chemical effects. These three effects can be identified broadly with types of nuclear radiation environments.

Displacement Effects . These cause permanent damage to crystal structures in semiconductors. They are caused by particle interactions with semiconductor-material nuclei, and are therefore produced by radiation environments, which either contain heavy particles in the flux or contain a large amount of high-energy photons. These high-energy photons produce secondary particles that create displacements. Radiation environments of this type are (1) nuclear reactors, (2) solar flares, (3) Van Allen belts, (4) cosmic radiation, (5) nuclear explosions, and (6) particle generators.

Transient or Ionization Effects. These are temporary effects in semiconductors. However, circuit failures can be experienced due to high induced currents if circuits are not carefully designed. As described in the section on types of radiation effects, the ionization effects are caused by electrons created via photon or other charged-particle interactions. The effects are proportional to dose rate or dose of the radiation pulse and are phenomena associated with electromagnetic (X- and gamma) and charged-particle radiation. Since semiconductor devices are not severely affected until the dose rate rises above 10^5 rads (C)/sec, the ionization effects are only experienced in radiation environments which produce these high dose rates. These radiation environments are (1) nuclear rocket engine start-up, (2) nuclear explosions, (3) pulsed nuclear reactors, (4) linear accelerators, (5) flash X-ray generators, and (6) close proximity to solar flares.

Surface Effects. Surface effects, first experienced as a practical design problem on the Telstar satellite, affect, as the name implies, the surface layers of transistor and semiconductor devices. After prolonged exposure to low-dose-rate electromagnetic (X- and gamma) radiation, gas ions and ions of foreign material formed near the surface of activated transistor devices are attracted into the surface layers of the semiconductor, thereby upsetting the internal electric fields. Doses of the order of 10^4 to 10^7 rads (C) are required to severely reduce the transistor performance. Any radiation environment which contains a large amount of ionizing radiation, either electromagnetic or charged-particle radiation, will produce the surface effects. These environments are (1) the Van Allen belt, (2) solar flares, (3) long-term exposure to nuclear reactors, and (4) long-term exposure to gamma radiation sources (for instance, Co^{60}).

Semiconductor Operation

In preparation for the sections below on the three radiation effects it may be well to describe the basic mechanisms at

work in semiconductor material prepared to form transistor
and diode junctions. Many excellent texts have been prepared
specifically on this subject and are listed in the bibliography
at the end of the chapter. However, in order to facilitate understanding of the effects physics, some salient points on semiconductor physics are described below.

Semiconductor Physics.   Current in a semiconductor can be
carried either by electrons with energies in the conduction
band or by holes (missing electrons) in the valence band.
Figure 2-2 shows the energy-level diagrams for a metal, an
insulator, an intrinsic semiconductor, a p- and n-type semiconductor.   Figure 2-3 shows the difference between a single
atom and one in a crystal structure [2]. The relative width of
the forbidden gap and the occupancies of the allowed energy
bands determine the material type. The Fermi level is defined
as the energy level at which the probability of occupancy is
50%. Metals are characterized by the fact that the valence band
is not completely filled and, because of the absence of a forbidden gap, the electrons can cause conduction at the slightest
provocation in the form of an electrical field. The insulator
on the other hand has a completely filled valence band with the
empty conduction band several electron volts away. Thermal

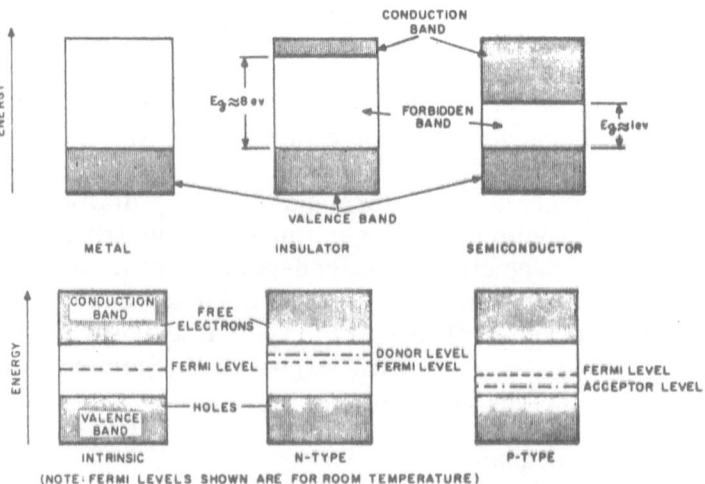

Fig. 2-2. Energy levels in semiconductors and solids.

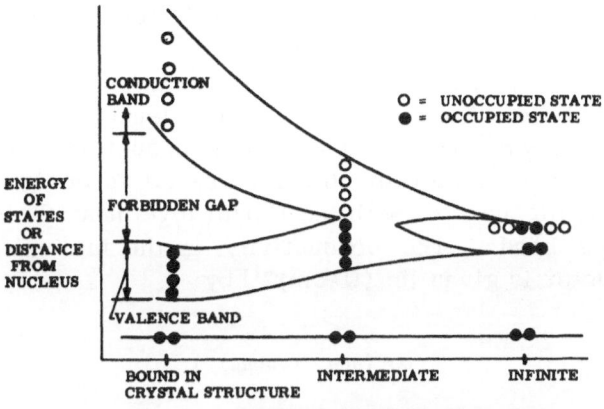

Fig. 2-3. Atomic electron distribution in a single atom vs. that of one which is bound in a crystal structure (after Dewitt and Rossoff, Transistor Electronics, McGraw-Hill Book Company, New York).

agitation at temperature is not sufficient to lift the valence electrons into the conduction band. No current will therefore flow under normal conditions in an insulator. However, if the insulator is heated to a high temperature or exposed to ionizing radiation, current will flow as electrons are made available in the conduction band.

The semiconductor, as the name implies, has characteristics which range somewhere between those of the metal and the insulator. Its forbidden gap is less than 1 eV and thermal agitation at room temperature does make valence electrons enter the conduction band. The forbidden gap in all cases described above assumes a pure crystal. In reality, impurities exist in gap, which either by donating or collecting electrons upsets the conductivity of the crystal. It is therefore possible to create p- or n-type semiconductor crystals by introducing small concentrations of impurities with the appropriate characteristics. Pentavalent (five free electrons) impurities such as arsenic, antimony, and phosphorus produce allowed electronic states near the edge of the conduction band. The fifth electron remaining when the impurity has joined in the crystal structure moves into the conduction band, and an n-type semiconductor is formed. The Fermi level then moves from the center of the forbidden gap toward the edge of the conduction band.   Trivalent (three free electrons) impurities such as boron, gallium, or indium gather near the edge of the valence band and pick up the fourth electron needed for a bond from the valence band. The deficiency in the valence band is called a hole (missing electron) and since the conduction is now primarily due to hole movements, the crystal is designated as a p-type semiconductor with the Fermi level near the edge of the valence band.   The conductivity in the intrinsic (pure) semiconductor is given [in $(\Omega\text{-cm})^{-1}$] by

$$\sigma = (ne\mu_n + pe\mu_p) \tag{2-1}$$

where n is electrons$/\text{cm}^3$ in conduction band, p is holes$/\text{cm}^3$ in valence band, e is electronic charge (coulombs), μ_n is electron mobility $(\text{cm}^2/\text{V-sec})$, and μ_p is hole mobility $(\text{cm}^2/\text{V-sec})$. For the extrinsic p-type material the $ne\mu_n$ term is smaller than the $pe\mu_p$ term, and is generally omitted.

Crystal Structure. To understand the concept of displacement threshold energy, E_d, to be used in the discussion on displacement effects, the makeup of the crystal structure will here be described first.

The semiconductor materials germanium and silicon, from which most transistors are made, belong to the carbon family. It may be intrinsic with a resistivity of from 45 to 60 Ω-cm, but for proper operation in a transistor is made extrinsic. Extrinsic material is achieved by adding impurities to the purified semiconductor material. As an example, one impurity atom per 10^8 germanium atoms reduces the resistivity from 60 to 4 Ω-cm. The crystal structure of the carbon family can be exemplified by the diamond lattice. In the above discussion of band theory, it was shown how the electron cloud surrounding a single atom differed from that of the electron cloud in the lattice atom.   It is seen how the valence band is dropped to states that are much closer to the inner shell. In the process energy was released in the form of heat. The amount of energy lost by the valence band electrons in the formation of the crystal is a measure of the binding force of the crystal structure. This energy E_c, called the energy of sublimation, is from

3 to 6 eV. Application of an energy equal to E_c is adequate to generate Frenkel pairs, defined as the existence of a vacancy and an interstitial atom that combine easily.

However, to remove an atom entirely and irreversibly from its lattice site in the crystal structure, an energy approximately four to five times greater than this binding force must be imparted to the lattice atom. In radiation effects terminology this is termed the displacement threshold energy E_d.

Conductivity in the semiconductor is a direct function of electron and hole concentrations and any change in these concentrations would affect the semiconductor device behavior. However, it takes a great deal of displacement damage before changes in conductivity manifest themselves. So much displacement damage is required that for all but very-high-frequency ($f_{a\,c\,o} > 500$ MHz) transistors and majority-carrier devices (tunnel diodes) another damage mechanism is used to monitor the displacement damage. This damage mechanism is the radiation-induced change in carrier lifetime τ. The carrier lifetime is defined as the relaxation time needed for an excess of electrons and holes to recombine. This carrier lifetime is a very important parameter in a transistor. The transistor as shown in Fig. 2-4 consists of two relatively large volumes of the same type of semiconductor material, the emitter and the collector, separated by the base, which is a very thin layer of the opposite type semiconductor material. With bias voltages applied as shown in the figure, conduction carriers will be injected from the emitter into the base with a velocity determined by the applied field. Once in the base volume of the transistor, the carriers are termed minority carriers. Approximately 98 to 99% of those injected enter the collector region by diffusing across the base, and an apparent current amplification has taken place. The base thickness and the minority carrier's lifetime in the base region determine in part the electrical characteristics of the transistor. The base thickness remains constant for moderate particle fluences, while the lifetime of the minority carriers is shortened by radiation-induced crystal-lattice defects, which act as trapping or recombination centers for the carriers. The crystal sur-

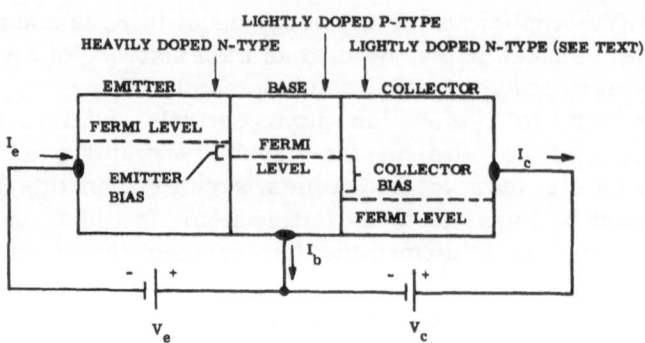

Fig. 2-4. Transistor action.

face constituting a discontinuity in the lattice also reduces the carrier lifetime, and changes in the crystal surface, such as those which occur due to ion trapping (see Surface Effects), will also influence the behavior of the device. As the amplification of the transistor is directly a function of the number of carriers traversing the base region, any loss of carriers via radiation-induced defects constitutes a loss in the device gain.

Lifetime, Recombination Centers, and Traps.   As discussed, the semiconductor material in its intrinsic form in the absence of any external electric fields or radiant energy will maintain an equilibrium condition with respect to carrier density. Diffusion of carriers will tend toward this condition.

Should this equilibrium be upset in a step-function fashion and the excitation immediately be removed, the induced disturbance will decay exponentially toward the original equilibrium condition. Mathematically this is shown as follows [2]:

$$n = n_i + \Delta n \ \exp(-t/\tau) \qquad (2-2)$$

where n is carriers/cm^3 at time t, n_i is the initial condition (carriers/cm^3), Δn is the induced carriers, and τ is the carrier lifetime.

It is this last parameter τ, the carrier lifetime or specifically the minority carrier lifetime, which is of great importance in radiation effect studies. The carrier lifetime is that length of time in which the induced equilibrium disturbance decays to 1/e of its original value. The process by which this takes place is termed r e c o m b i n a t i o n, and as the name suggests it takes place when a hole and an electron meet and recombine.

This direct recombination of a carrier dropping from the conduction band directly into the valence band is a major source of recombination in compound semiconductors such as GaAs and InAs, but is a very rare occurrence in germanium and silicon. The concept of the r e c o m b i n a t i o n   c e n t e r is required to account for experimentally measured values of lifetimes in germanium and silicon. The recombination center captures and removes from the free carrier stream the carriers that happen to pass. It holds the carrier until such time as an opposite carrier appears in the valence band below, whereupon the electron drops and combines with the hole in the valence band.

Thus the recombination center provides for permanent loss of the carrier. The recombination center is located at a defect, which creates an energy state in the middle of the forbidden gap. The defect may be located in the bulk volume of the crystal, such as impurity atoms or radiation-damage-induced lattice deformations. Of course, the surface of the crystal represents a defect extended enough to warrant the term s u r f a c e   r e c o m b i n a t i o n s, which accounts for irregularities in lifetime behavior at the crystal surfaces.

The t r a p, in contrast to the recombination center, removes the carrier only temporarily from the conduction band. The trap is produced by defects (impurity or radiation-induced) which create shallow localized states close to the edges of the forbidden gap. After capturing a carrier the trap will with much higher probability return the carrier to the conduction band than allow a recombination. The trap retains the carrier for periods ranging from 10^{-8} sec to several days before returning it to the conduction band, thereby introducing phase distortions for signals processed by the transistors [3].

Diffusion and Drift. The d i f f u s i o n   c o n s t a n t D is an important parameter in radiation-effect studies. Diffusion is one way in which carriers in the semiconductor move. The motion occurs without outside provocation in the form of an external electrical field, and is strictly analogous to the diffusion of gas molecules which are redistributing themselves according to temperature gradients.

Diffusion is motion perpetuated by the need for equilibrium. The carriers, electrons, and holes in a neutral semiconductor sample under the influence of room temperature diffuse to reach an equilibrium of equal carrier density throughout the crystal. The constant D associated with this diffusion, for intrinsic germanium, is equal to 100 cm^2/sec for electrons and 50 cm^2/sec for holes and, for intrinsic silicon, is equal to 31 cm^2/sec for electrons and 13 cm^2/sec for holes at room temperature. It is the ratio of the mean free path of the carriers squared to two times the mean free time. In mathematical terms [2] it is given as

$$D = \frac{\bar{l}^{\,2}}{2\bar{t}} = \frac{\bar{l}\bar{v}}{2} = \mu\frac{kT}{e} = \mu(0.026) \qquad (2\text{-}3)$$

where $\bar{l}$ is the mean free path in centimeters of diffusing carrier, $\bar{v}$ is the mean velocity (cm/sec), $\bar{t}$ is the mean free time (sec), μ is the mobility (cm^2/V-sec), and kT/e is the internal field (V) (0.026 V at room temperature) where k is the Boltzmann constant, T is temperature, and e is electric charge.

Drift of carriers in most regions of an operating transistor or other solid-state device dominates the diffusion of carriers. Drift is the motion of carriers due to electric field gradients. The carriers in the times between collisions with lattice elements come under the influence of the electric field and move continually in the direction dictated by this field.

Since the mass of the electrons is minute compared to the lattice-element mass, the electron will after a collision start with a completely random velocity to which the drift velocity will slowly add. One may then specify an average drift velocity, which is given by [2]

$$\bar{v}_d = \frac{e\bar{t}}{m}\,\epsilon$$

where e is the electric charge, $\bar{t}$ is the mean free time, m is the electron mass, ϵ is the electric field, and the term $e\bar{t}/m$ is the mobility μ. This is the term given in the conductivity equation (2-1) above, and is the parameter which is affected severely by large doses of particle radiation. It is a term which will often be encountered by the reader in data showing mobility variations vs. particle dose.

Diffusion Length. Diffusion length L is a parameter which neatly combines the two more important parameters of interest in radiation–effect studies of the semiconductor device, namely, carrier lifetime τ and carrier mobility μ. It is defined as that distance through which a group of carriers N may diffuse before their number has been reduced by $1/e$. Mathematically, it is presented [2] as follows:

$$L = \sqrt{\tau D} = \sqrt{\tau\mu\frac{kT}{e}} = \sqrt{0.026\,\tau\mu} \tag{2-4}$$

where L is in centimeters, τ in seconds, μ in centimeters squared per volt-second, and kT/e in volts, 0.026 V at room temperature.

Diffusion length can be calculated easily by measuring the hole current across the junction and inserting the following equation [2]:

$$L_p = \frac{b}{(1+b)^2}\left(\frac{kT}{e}\frac{\sigma_i{}^2}{\sigma_n}\frac{1}{I_{gp}}\right) \tag{2-5}$$

where L_p is diffusion length (cm) for minority carriers in n-type material, b is μ_n/μ_p (n refers to electrons; p refers to holes), k is the Boltzmann constant, σ_i is intrinsic conductivity, σ_n is doped conductivity of n-material, I_{gp} is hole current across the junction, and e is electronic charge.

Hall Measurement. The conductivity equation (2-1) above has two terms which are affected by displacement damage, namely, (1) carrier concentration (n or p) and (2) mobility (μ_n or μ_p). The reduction in the carrier concentration in the base region of a transistor is observed by monitoring the decrease in the minority carrier lifetime. Because this lifetime is a function of injection level (i.e., emitter-to-base current), it is important to specify the injection level. Direct measurement of the conductivity or its reciprocal, resistivity, will also show changes in the carrier concentration at moderate particle fluences [$< 10^{14}$ neutrons (> 10 keV)/cm^2]. At higher particle fluences, however, the mobility term is increasingly affected, and a monitoring technique which allows observation of mobility changes is required. A measurement which yields the Hall coefficient is used for this purpose. The measurement, because of its cumbersome nature, is performed on samples of n- or p-type semiconductor material rather than on transistor devices. An electrical current is passed through the sample, and the sample is inserted in an intense magnetic field. Voltage probes are attached to the sample perpendicular to the direction of both the current and the magnetic field. The Hall coefficient R is calculated by inserting the various measured parameters in the following equation:

$$R = \frac{V_H t}{HI} \tag{2-6}$$

where V_H is the Hall voltage, i.e., the potential developed in a direction perpendicular to the magnetic field direction; t is the thickness of the sample in magnetic field direction; H is the magnetic field; and I is the current through the sample in a direction perpendicular to the direction of H and V_H. With R available, the actual carrier concentration in extrinsic samples may be calculated from

$$n = \frac{-r}{Re} \quad \text{(for n-type material)}$$

$$\tag{2-7}$$

$$p = \frac{r}{Re} \quad \text{(for p-type material)}$$

where e is electronic charge and r is a constant (μ_H/μ) ranging from 1 to 2. μ_H is termed the Hall mobility and is the product of R, the Hall coefficient, and σ, the conductivity. With R and σ measured the displacement damage induced changes in mobility, μ, can be calculated.

Displacement Effects

As pointed out above, the effects of radiation on the lifetime of minority carriers, carrier concentration, and mobility are the most important factors affecting the performance of transistors. It has been found [4] that much of the degradation at low integrated doses in transistors is caused primarily by interactions between the radiation-induced defects and the impurity atoms already in the crystal. Further experimental evidence is available [5] showing that a vacant lattice site can move about in the lattice at room temperature. The vacancy will diffuse through the lattice until trapped to form defects which are stable at the sample temperature. Vacancies will often capture a displaced atom, thereby eliminating the defect. However, as the total dose increases and the damage is caused by heavy, high-energy particles, regions of high defect density are formed with linear dimensions of up to 200 Å; these do not anneal as readily as those formed with lower doses.

The following words are part of the terminology of displacement effects:

Vacancies: Empty places in the crystal lattice created as the atom is removed by a particle collision.

Interstitial atoms: Atoms which after having spent the imparted collision energy have become lodged in interstitial positions in the lattice.

Impurity atoms: Original crystal atoms, which have been transmuted by a nuclear reaction (not to be confused with doping agent atoms).   An impurity atom can also be formed when the displacing particle is captured in the crystal lattice.

Thermal spike: A primary recoil or knock-on may under certain circumstances lose all its energy in a small volume containing only about 10^4 atoms. When this happens, high temperatures in the excess of 1000°K are generated in less than 10^{-11} sec. This occurrence is termed a thermal spike [7].

Displacement spike: The local disturbance created by a primary which causes many secondary displacements.

Displacement Effects Versus the Primary Particle.   Neutrons, having no electrical charge, make direct hardsphere collisions with nuclei, and all energy transfers between the Wigner energy threshold (E_d) of 10–30 eV up to the maximum are possible. The maximum is given by

$$E_p = \frac{M_1 M_2}{(M_1 + M_2)^2} \left(\frac{E_K}{E_d} \right) \tag{2-8}$$

where E_p is the energy of primary knock-on, M_1 is the mass of the neutron, M_2 is the mass of the target, E_K is the kinetic energy of the neutron, and E_d is the Wigner energy.

An average number of between 317 and 2500 displacements are created per collision by fission neutrons (> 100 keV). Neutrons produce these displacements in a small spherical region called a cluster. The vacancies resulting are mainly trapped in a highly disordered region of the cluster. A neutron-produced displacement cascade therefore consists of a disordered region with various lattice sites empty. No reliable models have been developed relating physical properties of the cascade clusters in terms of size, shape, distribution, recombination, and trapping.

For protons with energies below 500 MeV, the collisions produce displacements by Rutherford scattering in which the Coulomb charge field of the proton and the nucleus interact to permit exchange of momentum and energy. The mean number of atoms displaced per centimeter by a proton is given by the following equations:

$$\eta_D(E) = \frac{964}{E_K} \ (9.43 + \ln E_K) \quad \text{for silicon} \tag{2-9}$$

$$\eta_D(E) = \frac{735}{E_K} \ (7.66 + \ln E_K) \quad \text{for germanium} \tag{2-10}$$

where E_K is the kinetic energy of the proton.

In contrast to the neutron, the proton produces only small displacement cascades of the order of three to eight per proton collision. There is a good probability that these vacancies will escape the damage center and move by thermal diffusion into the undamaged lattice and become trapped by impurity centers. The difference between neutron and proton damage is that neutron damage effects are more dependent on the size of the cluster than on the type of defect in the cluster. Changes in impurity concentrations or doping will have little effect on the total damage. Proton damage on the other hand is strongly dependent on impurity concentrations and the type of impurity.

Electrons must travel at relativistic velocities in order to displace atoms. Again, as the electron is a charged particle, the dominant mechanism is Rutherford scattering. The maximum energy that can be transferred in a collision is given by

$$E = \frac{2(E_K + m_0 c^2) E_K}{M_1 c^2} \qquad (2\text{-}11)$$

where E_K is the kinetic energy of the electron, c is the velocity of light, m_0 is the electron rest mass, and M_1 is the mass of the target particle.

The displacement per centimeter is calculated from the following equation:

$$\eta_D\ (E)\ =\ N_a\ [\sigma_D(E)]\ [\bar{V}\ (E)] \qquad (2\text{-}12)$$

where N_a is lattice atoms/cm^3, σ_D is the displacement cross section, and $\bar{V}$ is the mean number of secondary displacements per primary recoil. This equation is shown in graph form for silicon and germanium in Fig. 2-5.

The bombarding particle makes one or a number of primary collisions. The knock-on atom, the primary recoil or what is also known as the primary, goes on to produce more displacements on its own, and these displaced atoms may in turn produce what is known as tertiary displacements. This multiply-

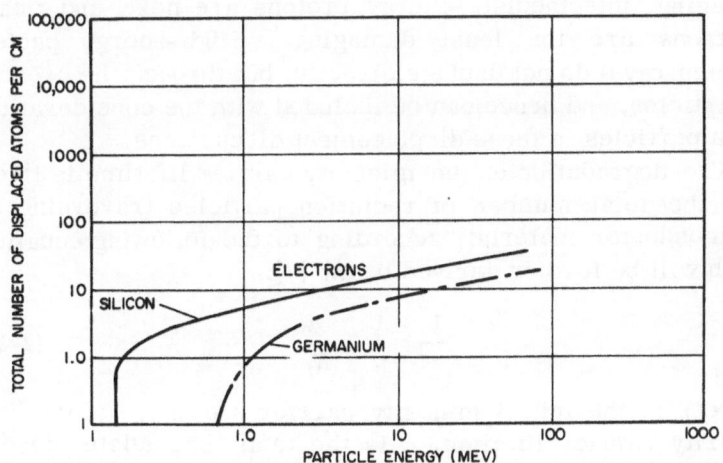

Fig. 2-5. For electrons, total number of displacements vs. electron energy.

ing effect leads to the displacement cascade which accounts for the major portion of the defects created by heavy particles.

The defects in materials are termed either dislocations or point defects. The dislocation is a linear defect around which the lattice is disturbed for only a few interatomic distances radial to the dislocation line. The point defect is the term associated with lattice vacancies, interstitial atoms, and impurity atoms. The lattice is in this case disturbed for several interatomic distances around the point defect.

Probably more important are the composite defects which are created from the elementary defects. Some of these are as follows [6]:

1. Creation of two adjacent vacancies which may produce a defect that has quite different diffusive characteristics than a single vacancy.
2. Large concentrations of vacancies leading to volume imperfections.
3. The creation of a closely spaced "Frenkel pair;" this condition is typically produced by electron bombardment.

The Energy of the Particle. As just described, the particle type, whether neutron, proton, or electron, is important in determining the extent and severity of damage to semiconductors. When a collision takes place, fast neutrons are the most damaging, intermediate-energy protons are next, and finally electrons are the least damaging. High-energy photons (gamma rays) do not damage directly, but through the creation of particles, and hence can be included with the considerations given particles in these displacement discussions.

The degradation of the minority carrier lifetime is linear with the total number of radiation particles traversing the semiconductor material, according to the following equation, which will be further derived below [7]:

$$\frac{1}{\tau} = \frac{1}{\tau_0} + \frac{\phi}{K_\tau} \qquad (2\text{-}13)$$

where τ_0 is the initial minority carrier lifetime, τ is the final minority carrier lifetime, ϕ is the total accumulated dose in particles/cm^2, K_τ is the damage constant which is experimentally determined and adjusts for variables such as particle spectrum, particle type, and device construction.

The spectrum of the particle field is extremely important to include, since the radiation damage is not linear with particle energy. To account for this fact, the energy spectrum of the particle field and the accuracy with which it is quoted should always be ascertained.

The damage or displacements vs. energy will now be examined. Electrons cannot produce displacements until their kinetic energy exceeds at least 200 keV. The displacement probability for greater energies fall somewhere between the $\frac{1}{2}$ and $\frac{1}{3}$ power of the kinetic energy.

For protons, the threshold energy for displacement damage is 13 eV for silicon and 31 eV for germanium. The displacement damage as a function of proton energy from that energy up follows approximately the inverse of the proton kinetic energy.

For neutrons the threshold energy for displacement damage is approximately 200 to 500 eV. Above this energy the displacement damage as a function of energy is about linear to 0.2–1 MeV and from that point follows a 0.5–0.75 power of the energy.

These energy vs. damage effects curves are of great help when comparing two spectra of the same type of particles. By normalizing each spectrum to one particular energy, the induced damage can be compared much better.

Displacement Effects on Various Semiconductor Devices

The operational semiconductor devices that are available today achieve their performance characteristics by utilizing extrinsic semiconductor materials. These are intrinsic materials doped by the use of appropriate impurities to become either n-type or p-type extrinsic semiconductor materials. As described above, this was accomplished by adding impurities to the intrinsic materials with such properties that localized energy states were created in the forbidden band just below the conduction band in the case of an n-type extrinsic material, and just above the valence band in the case of a p-type extrinsic material.

The level or amount of impurities added plays an important role in determining the characteristics of the final semiconductor device, as does the combination of the n - and p-type extrinsic materials. As an example, consider a high-frequency npn transistor. The emitter is heavily doped. It has a high percentage of impurities in order to make a large number of electrons available for the eventual conduction. The base consists of a p-type extrinsic material. However, it is relatively lightly doped to minimize direct recombinations as the electrons during the conduction enter the base region. Furthermore, the base region often has a graded impurity density to enhance high-frequency operation. The concentration of impurities is higher at the emitter junction than at the collector junction. The collector is an n-type extrinsic material, but in contrast to the emitter is only lightly doped. The collector region impurity concentration is based on current and voltage requirements. High breakdown voltage requires a low impurity concentration, whereas high current requires high impurity concentration.

The careful balance of impurity concentrations then determines the operational characteristics of any semiconductor device. Therefore the unbalance introduced by radiation-induced defects affects the device performance characteristics. As an example of these effects, conductivity changes lead to changes in breakdown voltages of transistors. Since radiation-induced defects act much like impurity-induced defects when a low dose of the right type of particles is used, diodes can be made to switch faster by irradiation with electrons. Indeed, this method is used by diode manufacturers.

To emphasize that most radiation-induced defects are permanent, a few words may be added about annealing. Annealing of defects occurs when a vacancy and an interstitial atom combine. For less massive destruction of the crystal structure, such as that produced by electrons and protons, annealing of a certain number of created defects will occur at room temperature. However, for neutron-induced defects the semiconductor material must be heated to temperatures in excess of 150°C to produce significant annealing.

In general then the radiation-induced defects can be considered permanent damage. A very slow annealing is experienced when irradiated devices are permitted to rest for several months at room temperature. However, the major fraction of the defects induced during the irradiation will remain.

Transistors. The degradation of the transistor performance may qualitatively be described in the following manner. The current flowing in the collector circuit divided by the current flowing in the emitter circuit of a common base transistor circuit with commonly applied bias potentials is given by

$$\frac{I_c}{I_e} = \frac{1}{1 + (2W/L)^2} \tag{2-14}$$

where W is the basewidth and L is the diffusion length of minority carriers in the base, given as $\sqrt{\tau\mu(kT/e)}$ in equation (2-4). The equation is the slope of the curve showing emitter current vs. collector current in the transistor. For incremental changes of these currents it is termed the short-circuit gain α. The changes of α with particle dose are less perceptible than the changes in β or h_{FE}, the forward current transfer ratio which is related to α by the equation

$$\beta = \frac{\alpha}{1 - \alpha}$$

From these equations it is clear that the transistor operation is affected by changes both in minority carrier lifetime and mobility, since both determine L. The term β or h_{FE} is more frequently encountered since it relates to common emitter circuits. It also shows in better detail than α the changes to the transistor with particle dose.

For the purpose of the present discussion, transistors may be divided into two broad categories: low-frequency devices and high-frequency devices. The low-frequency devices are characterized by relatively long minority carrier lifetimes (up to 1 msec) whereas the high-frequency devices have very short lifetimes (as small as a fraction of a nanosecond). The onset of degradation therefore occurs in the low-frequency devices long before high-frequency devices are affected. Once initiated, the degradation can be calculated by equation (2-17), given below.

In this region of the degradation curve, which we shall term the minority lifetime degradation region, both the low- and high-frequency devices degrade approximately linearly with dose. The low-frequency devices quite soon are degraded so severely that they are no longer useful. The high-frequency devices on the other hand with a short lifetime already built in continue to be useful beyond the minority lifetime degradation region.

The high-frequency devices then enter the second region, which is separated from the first by a knee in the degradation curve. This region we shall term the conductivity degradation region, as it is now the change in both carrier concentration and mobility that dominates the device degradation.

The change in performance with dose in this region is slow until a complete loss of performance is encountered as the semiconductor material starts to change its physical crystal structure characteristics. This last region may be termed the device threshold. The sequence of regions is depicted in Fig. 2-6.

The experimental work over the years has been concerned with creating theoretical and empirical expressions, which

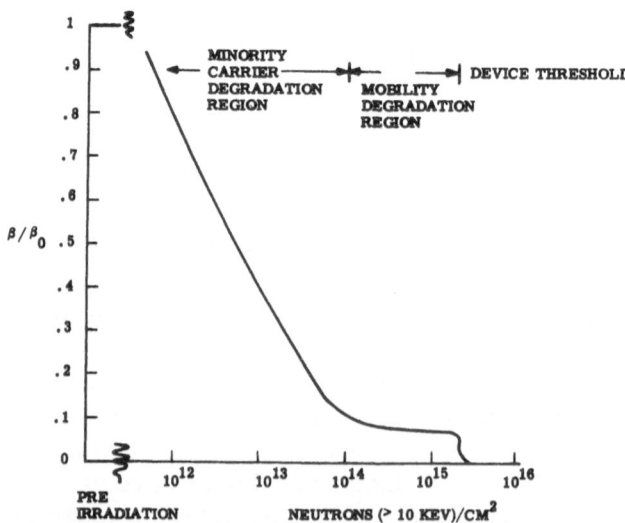

Fig. 2-6. Typical transistor h_{FE} or β degradation sequence. (The neutron spectrum in this typical case would be the 1 MeV average spectrum of the TRIGA MkI reactor.) See also Fig. 2-32.

would allow prediction of device behavior when exposed to particle irradiation. Early in 1957 J. W. Easley [8] introduced the concept of a damage constant, which could be used to account for differences in device construction and particle type. J. J. Loferski [7] and G. C. Messenger and J. P. Spratt [9] expanded on the concept, and it became known as the lifetime degradation constant K_τ defined by the expression

$$\frac{1}{\tau} = \frac{1}{\tau_0} + \frac{\phi}{K_\tau} \qquad (2\text{-}15)$$

where, as given earlier, τ is the final minority carrier lifetime, τ_0 is the initial minority carrier lifetime, and ϕ is the total integrated particle dose. J. R. Bilinski [10], using J. J. Loferski's expression for K_τ, which is given as

$$K_\tau = \frac{f_{aco}(\beta_0/\beta - 1)}{0.194 \, \phi \, \beta_0} \qquad (2\text{-}16)$$

where f_{aco} is the alpha cutoff frequency of the transistor, β_0 is initial β, β is final β, and ϕ is the total integrated particle dose, rearranged the terms as follows

$$\frac{\beta}{\beta_0} = \frac{1}{1 + (0.194 \, \phi \, \beta_0/K_\tau f_{aco})} \qquad (2\text{-}17)$$

and created the two nomographs shown in Figs. 2–7 and 2–8.

The reader should use the nomographs cautiously, since the damage constant K_τ shows wide variations from device to device as pointed out by J. R. Bilinski during his investigations into proton–neutron damage correlations [11]. The determination of K_τ and its limitations are described below. At this point it is sufficient to say that a particular K_τ is only usable in the degradation region for which it was determined. Since early work was mainly concerned with current gain degradation in relatively low-frequency devices, the nomographs should only be applied in the minority carrier lifetime degradation region. A predicted accuracy of 30–50% can then be expected.

The value K_τ is then an empirical constant which was determined for the base material of the transistor. Typical values are given in Table 2-I, which correctly shows the better performance, by one order of magnitude, that is characteristic

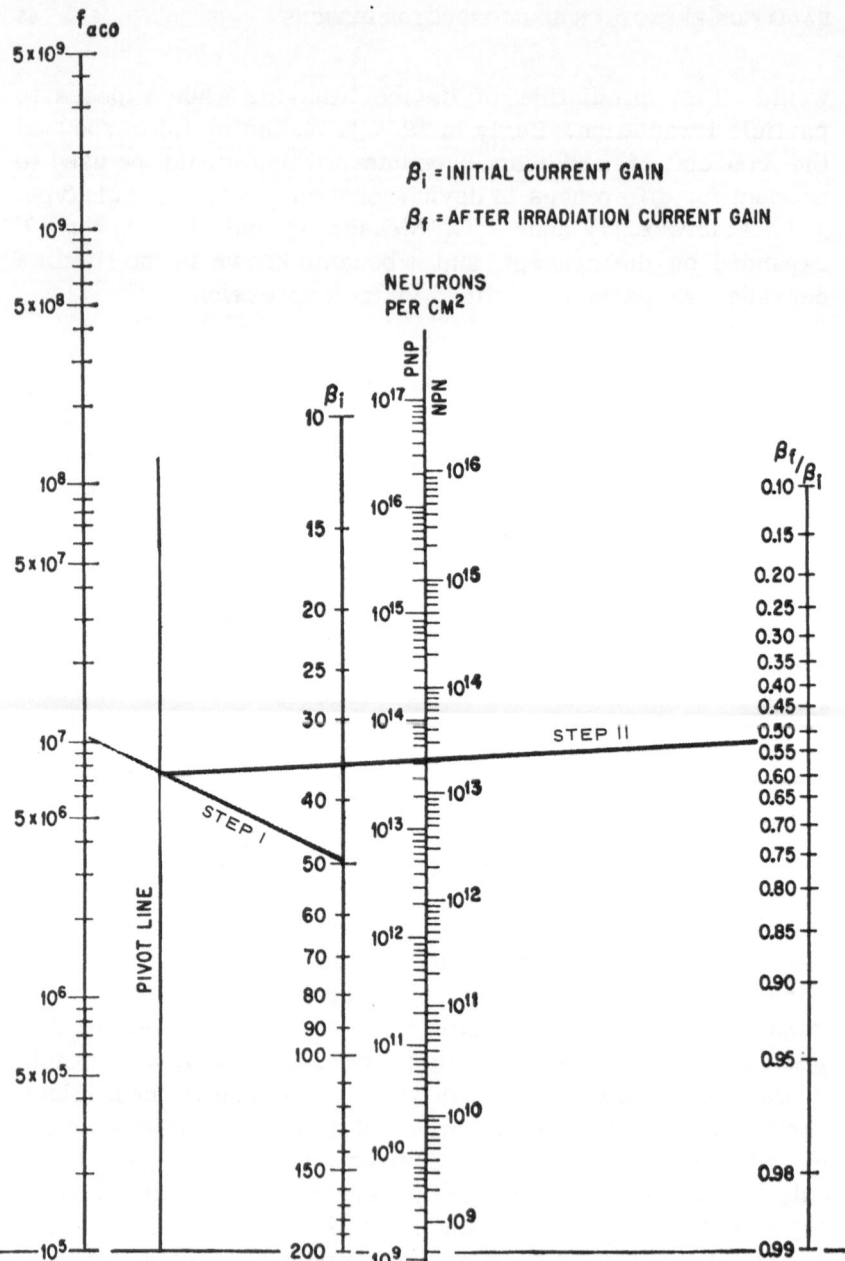

Fig. 2-7. Germanium transistor nomograph. To determine relation of neutron flux to current gain of a germanium transistor, if either value is known, connect $f_{\alpha co}$ (alpha cutoff frequency) with rated β_I. Line intersects pivot line. If neutron flux is known, draw a second line from intersection through this known value to β_f/β_I scale and read off current gain.

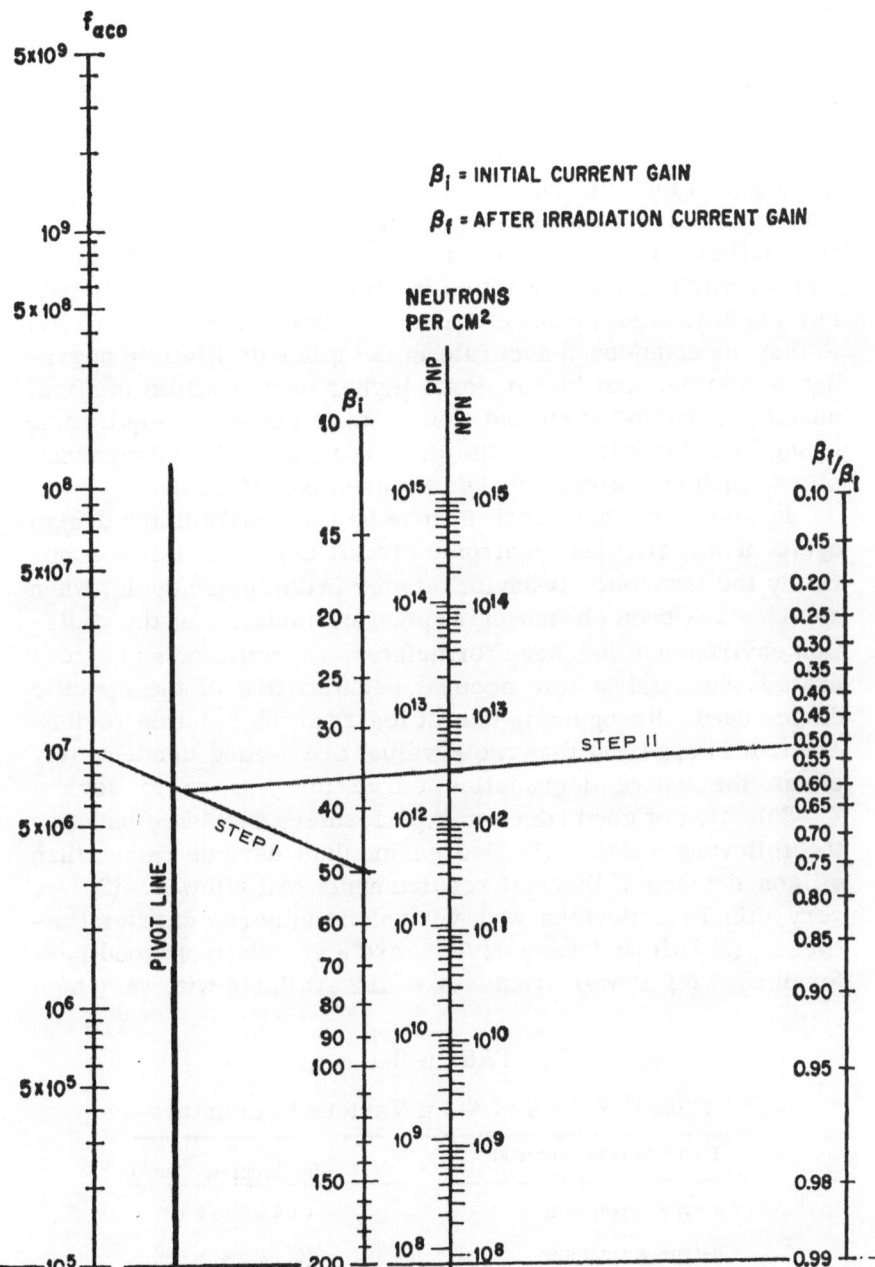

Fig. 2-8. Silicon transistor nomograph shows how grounded emitter current gain of silicon transistors varies as neutron flux varies.

of germanium transistors. However, as would be expected, such typical values cannot account for the many different doping techniques that a transistor manufacturer uses to obtain just the right transistor characteristics. Neither does K_τ account for surface combination and emitter efficiency changes due to irradiation. L. B. Gardner and A. B. Kaufman [12] point out that, as surface recombination is altered only in the early part of irradiation and emitter efficiency is a function of conductivity, K_τ may be considered accurate in the minority lifetime degradation region, that is, for doses higher than required to affect surface recombination but lower than the dose required to change conductivity. As the dose is increased, other parameters, such as voltage breakdown, are also affected.

In summary, a typical K_τ may be used early in the design cycle of a particular electronic circuit to predict fairly accurately the transistor behavior. Later in the design cycle, when a device has been chosen and a specific simulation of the radiation environment has been formulated, a specific K_τ can be generated which takes into account peculiarities of the specific device used. Recognizing that at least two degradation regions exist, it is apparent that two K_τ values are needed to adequately cover the entire degradation curve for a specific device.

Selection of good transistor performers should be based on the following rules: (1) Use germanium devices rather than silicon devices if thermal requirements will allow it. (2) Use very thin base devices with very short minority carrier lifetimes. (3) Diffused-base devices are also relatively good performers. (4) Power transistors are available with very high

TABLE 2-I

Typical Values of K_τ for Various Transistors

Transitor base material	K_τ (neutrons-sec/cm^2)
n-type germanium	$(5.0 \pm 2.0) \times 10^7$
p-type germanium	$(2.4 \pm 0.4) \times 10^7$
n-type silicon	$(2.8 \pm 0.8) \times 10^6$
p-type silicon	$(3.2 \pm 1.1) \times 10^6$

alpha-cutoff frequencies.   These make it possible to build
nuclear-hardened power inverters by using circuits which will
still operate satisfactorily after the device gain has dropped
to one.

   Selection of Transistors.   The selection of transistors for
operation in a radiation environment can be based on electrical measurements of certain parameters which are indica-
tive of the nuclear hardness of the transistor.   Such a technique was developed and substantiated by irradiation test data.
D. Hendershott [13] of General Electric developed the technique
from the relationship between base transit time and base width
as well as base doping.

   The base transit time is measured by using the technique
suggested by Lindmayer and Wrigley [14].   By plotting the reciprocal of gain bandwidth product as a function of the reciprocal of emitter current and extending this curve to cross at the
point where the reciprocal of emitter current is equal to zero,
one may read the base transit time.   Figure 2-9 shows such
a plot of 12 samples of 2N918 transistors.

   The technique as shown in Table 2-II was found to be quite
accurate, but only for transistors degraded into the conductivity
degradation region.   Three selection criteria for transistors are
used: (1) Using equation (2-17) or nomographs estimate the
post-irradiation gain.   Then select only those which will yield
acceptable gain.   This amounts to selection of transistors with
a initial minimum h_{FE} or β.   (2) Measure base transit time and
select only those transistors with a value below a predetermined
minimum value.   (3) Eliminate those devices which show an
abnormally steep slope of the $1/\omega - 1/I_e$ curve when compared
to other samples. (Samples 3, 6, and 10 in Fig. 2-9 are examples of this behavior.)

   Depletion-Region-Controlled Devices.   This type of device is
represented by the unipolar, the field effect, and the MOS
(metal oxide semiconductor) transistor.   They all work on the
principle of an electrically controlled depletion width.   Figure
2-10 shows the device in cross section.   By application of a
reverse bias across the gate, the depletion width in channel
is increased, thereby squeezing and decreasing the current
flow between the source and the drain.   A signal source be-

TABLE 2-II

Predicted Order of h_{FE} and Post-Irradiation Order of h_{FE} for Selected Standard GE 2N918 Transistors (reprinted with the permission of the General Electric Company, from D. Hendershott et al., General Electric Report ETR-8234-008)

Predicted order of h_{FE}		Post-test order of h_{FE}*
Sample number	Base transit time (nsec)	Sample number
7	0.102	33
34	0.108	7
2	0.112	2
38	0.115	34
19	0.115	38
1	0.108	25
33	0.122	19
25	0.116	26
24	0.124	12
26	0.126	1
16	0.121	24
12	0.130	16
10	0.138	6
6	0.144	10
4	0.162	4
Sample number	Pre-test h_{FE}	Post-test h_{FE}*
33	77.5	4.87
7	68.8	4.38
2	75.0	4.37
34	87.5	4.37
38	68.8	4.25
25	72.5	4.25
19	102.5	4.0
26	63.8	3.88
12	70.0	3.75
1	52.5	3.63
24	61.3	3.37
16	56.3	3.25
6	51.2	3.0
10	62.5	2.88
4	47.5	2.25

*After neutron dose of 2.0×10^{15} neutrons/cm^2 > 10 keV in General Atomic TRIGA MKI reactor, F-ring.

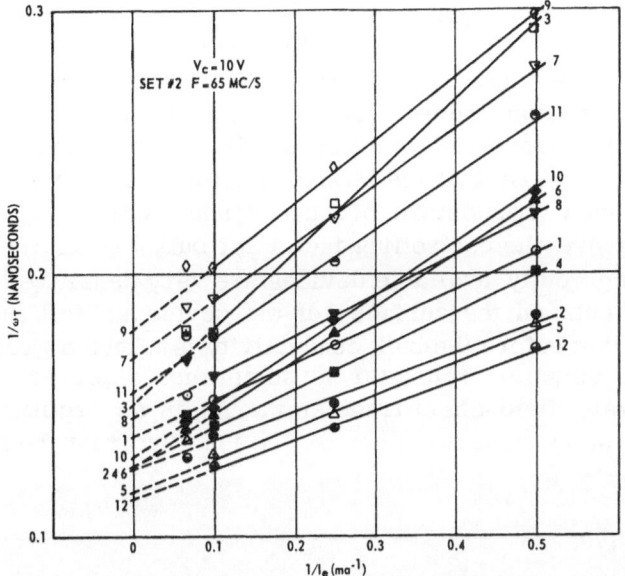

Fig. 2-9. $1/\omega_T$ vs. $1/I_e$, random selected 2N918 transistors (reproduced with the permission of General Electric, from D. Hendershott et al., General Electric Report ETR-8234-008).

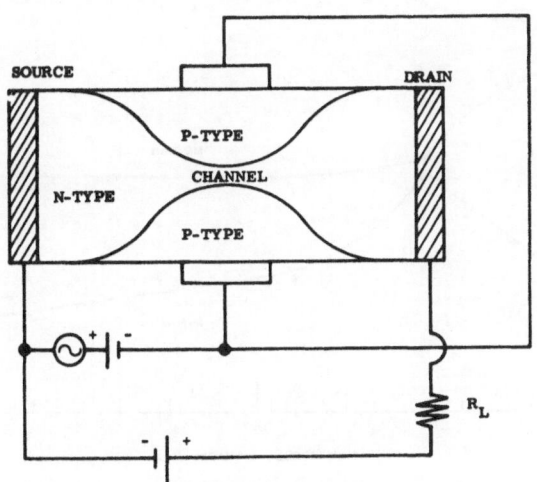

Fig. 2-10. Field-effect transistor (diode-gate type).

tween the source and the gate can then control a much larger current between the source and the drain.

All three devices fall from the standpoint of radiation resistance somewhere between high-frequency silicon and germanium transistors. From a bulk damage standpoint it can be shown clearly that the MOS transistor especially is superior to ordinary transistors because conductivity changes in the channel are the controlling factor for bulk degradation. However, currently available devices are very sensitive to radiation effects on the surfaces of the device and fail due to this cause long before the bulk conductivity is affected greatly.

The unipolar transistor is sometimes referred to as the diode-gate field-effect transistor. As in an ordinary diode, a reverse current flows across the junction from the gate into

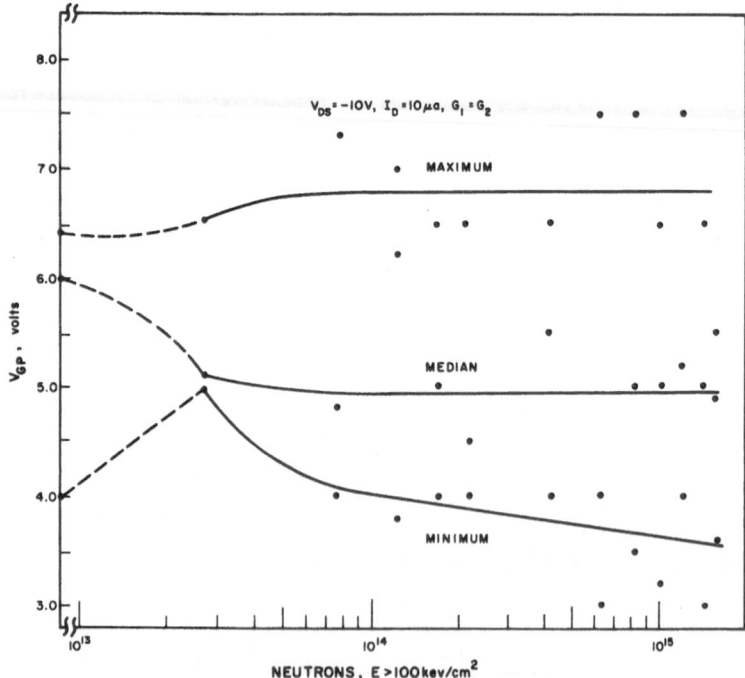

Fig. 2-11. Typical behavior of field-effect transistor pinch-off voltage with neutron dose (reproduced with the permission of the Texas Instrument Semiconductor Department, from L. Taylor, Texas Instrument Report, March, 1962).

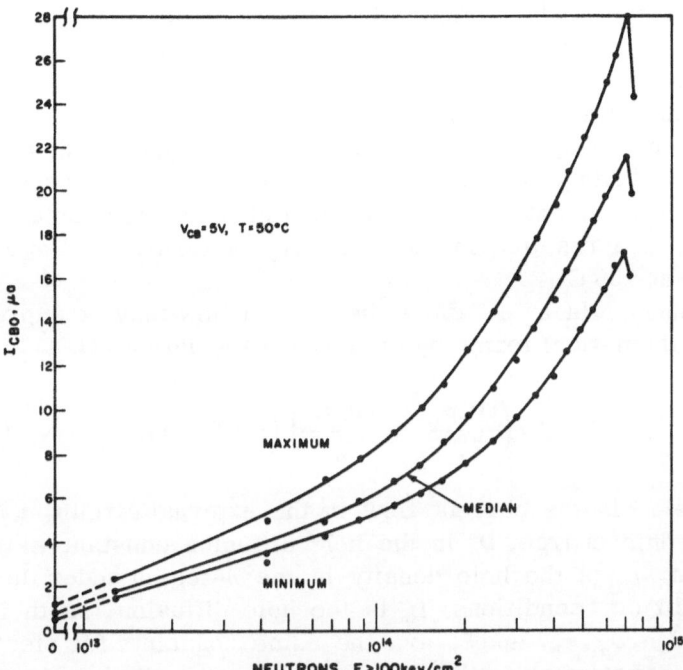

Fig. 2-12. The influence of neutron radiation the reverse collector current of a 2N705 transistor (GTR spectrum) (reproduced with the permission of the Texas Instrument Semiconductor Department, from L. Taylor, Texas Instrument Report, March, 1962).

the channel. As in the diode this reverse current increases with radiation dose. The reverse current is related to the pinch-off voltage, which therefore changes with radiation dose. This is shown in Fig. 2-11.

This reverse current is reduced manyfold in the MOS transistor, also called the insulated-gate field-effect transistor. In this device the channel is separated from the gate by an intrinsic layer of semiconductor material and it appeared from a theoretical standpoint that its radiation performance indeed would be strictly a function of conductivity changes in the channel. However, as was the case for transistors described in the section on surface effects, the changes in surface-state density in currently available MOS devices is the dominant damage mechanism [15]. Additional radiation exposure com-

pounds the surface effects decrease in device performance by
decreased substrate resistivity and carrier mobility in the
channel.

The increase of positive charges in the intrinsic layer
between the gate and the channel causes an increase in the
turn-on voltage, as shown in Fig. 2-30 from G.C. Messenger's
experimental data [15]. Annealing of the surface defects was
tried, and resulted in 90% recovery of turn-on voltage after
16 hr at 150°C.

Diodes. In a *pn* diode the current flow may be expressed
in mathematical terms by the following equation [2]:

$$I = \left(\frac{eD_p p_0}{L_p} + \frac{eD_n n_0}{L_n} \right) \left(e^{eV/kT} - 1 \right) \tag{2-18}$$

where I is the current flow in the external circuit, e is the
electronic charge, D_p is the hole diffusion constant in the n-
region, p_0 is the hole density in the n-region under thermal
equilibrium conditions, L_p is the hole diffusion length in the
n-region, D_n, n_0, and L_n are the same constants for electrons
in the p-region, V is the electric field across the junction, i.e.,
applied field minus diode IR voltage drop, and $kT/e = 0.026$ V
under room-temperature conditions.

The first term of the equation is the reverse saturation
current, I_s. To account for the space-charge region, terms
should be added to the expression for I_s (see Chapter 5 in
Beam [3]). Then, as is the case in transistors, the changes in
minority carrier lifetime will dominate the first term of the
equation at relatively low particle doses. As the lifetime decreases, the diffusion length decreases, and the reverse or
back bias current increases. These are indeed the early
effects in diodes. For switching diodes which are doped with
gold to purposely decrease carrier lifetime, the radiation-
induced lifetime change is hardly noticeable. Eventually at
larger integrated particle doses the conductivity begins to
change, thereby affecting the IR drop across the diode. This
influences the V in the exponential term, and the diode starts
to fail rapidly due to increased forward resistance. Figure 2-13
shows a typical diode performance vs. radiation dose curve.

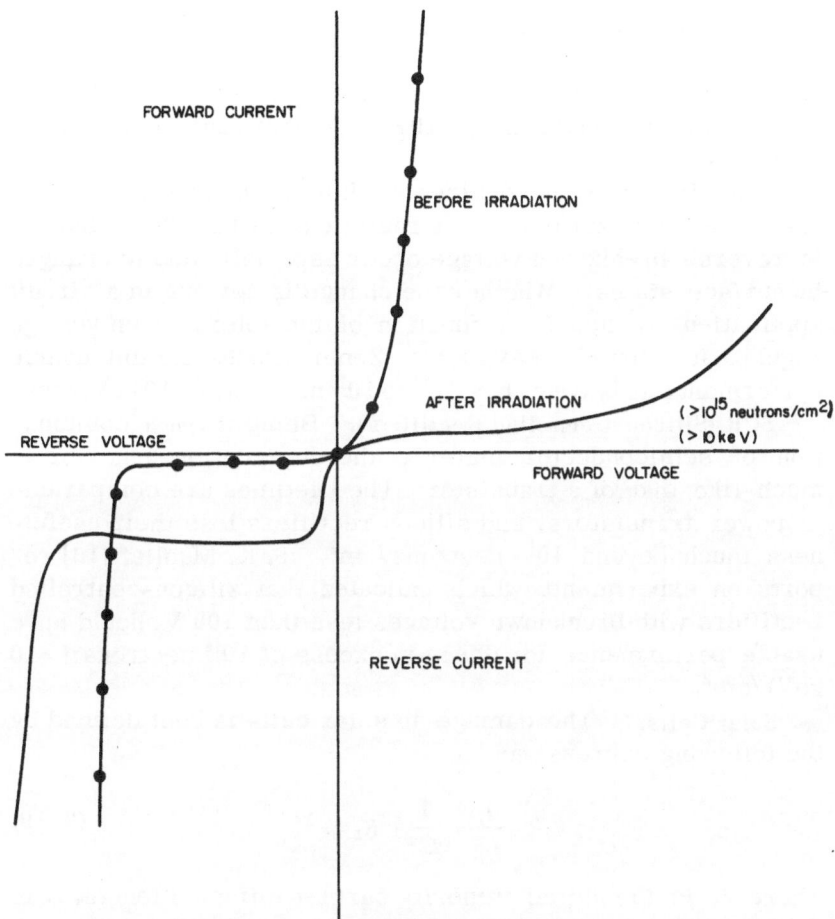

Fig. 2-13. Typical diode degradation.

It is apparent that the diode application determines which parameter change is important.

Rectifier Diodes. These are generally large-volume devices in order to increase power handling capability and to increase the breakdown voltage. Accordingly, relatively long carrier lifetimes are required. The major effect leading to failure of these devices is the increased reverse current with particle dose. Threshold dose is 10^{14} to 10^{15} neutrons (>10 keV)/cm^2 fission spectrum or equivalent dose of other particles.

Switching Diodes. These are very nuclear-hard as described above because of their inherently short carrier lifetime. Conductivity changes of consequence are not serious until doses in excess of 10^{15} neutrons ($> 10\,\text{keV}$)/cm^2 or equivalent of other particles.

Zener Diodes. These devices working in the avalanche mode are quite resistant to nuclear radiation. In Fig. 2-13 changes in reverse breakdown voltage occur especially due to changes in surface states. Whether the change is serious in a circuit application or not is a function of the tolerance on voltage regulation allowed. Available Zener diodes exhibit usable performance to between 5×10^{13} to 10^{15} neutrons ($> 10\,\text{keV}$)/cm^2.

SCR (Silicon-Controlled Rectifiers). Being a *pnpn* combination of semiconductor material the degradation behavior is much like that of a transistor. The lifetimes are comparable to power transistors, and silicon rectifiers lose their usefulness much beyond 10^{13} neutrons/cm^2. S. K. Manlief [16] reports on experiments which indicated that silicon-controlled rectifiers with breakdown voltages less than 100 V should have usable performance for doses in excess of 10^{14} neutrons (> 10 keV)/cm^2.

Solar Cells.   The damage in solar cells is best defined by the following expression:

$$\frac{1}{L^2} = \frac{1}{L_0^2} + K_L\,\phi \qquad\qquad (2\text{-}19)$$

where L_0 is the initial minority carrier diffusion length, L is the final diffusion length, $K_L = K_r/D$ (termed the diffusion length damage constant), and D is the diffusion constant. Table 2-III lists typical values for K_L for various environments [11].

Quantum Tunneling Device.   The tunnel diode achieves its operational characteristics by utilizing very heavily doped p- and n-type material on each side of the junction. Then at a low forward voltage which is less than the forward breakdown voltage, tunneling of carriers across the potential barrier occurs, giving rise to a negative slope of the V_t vs. I_t curve.

The heavy doping immediately tells us that the minority carrier lifetime is extremely short and therefore affected very little by particle dose. The tunnel diode is quite resistant to

TABLE 2-III

Summary of Diffusion Length Damage Constants for Proton and Neutron Irradiated Silicon Solar Cells (reprinted with the Permission of IEEE, from J. R. Bilinski et al., IEEE, NS-10, November 1963)

Cell type	Particle	Energy (MeV)	Damage constant, K_L (particles)$^{-1}$		
			Minimum	Average	Maximum
pn	Proton	96.5	5.8×10^{-6}	8.3×10^{-6}	11×10^{-6}
pn	Proton	68.9	9.2×10^{-6}	11×10^{-6}	13×10^{-6}
pn	Proton	48.5	17×10^{-6}	22×10^{-6}	25×10^{-6}
pn	Neutron	Fission spectrum	4.9×10^{-6}	6.1×10^{-6}	8.3×10^{-6}
pn	Neutron	Moderated spectrum	3.8×10^{-6}	4.6×10^{-6}	6.0×10^{-6}
np	Proton	96.5	1.2×10^{-6}	1.5×10^{-6}	1.8×10^{-6}
np	Proton	68.9	2.0×10^{-6}	2.4×10^{-6}	2.9×10^{-6}
np	Proton	48.5	2.4×10^{-6}	3.1×10^{-6}	3.8×10^{-6}
np*	Neutron	Fission spectrum	1.4×10^{-6}	1.5×10^{-6}	1.7×10^{-6}
np†	Neutron	Moderated spectrum	0.69×10^{-6}	0.92×10^{-6}	1.4×10^{-6}

*One cell with a K_L of 4.1×10^{-6} is not included in table.
†Two cells with K_L's of 5.2×10^{-6} and 7.3×10^{-6}, respectively, are not included in table.

particle dose as shown in Fig. 2-14. The change in peak-to-valley current ratio vs. particle dose is shown here.

The selection of tunnel diodes should be made remembering that the radiation tolerance is a function of current density, which in turn is related to the ratio of capacitance to peak current. Hence maximum peak current with a minimum capacitance signifies a nuclear-hard tunnel diode. Devices are are available which will withstand up to 2×10^{16} neutrons/cm^2 (> 10 keV).

Special Devices. There are other semiconductor devices not mentioned here, and the many that will undoubtedly appear in the

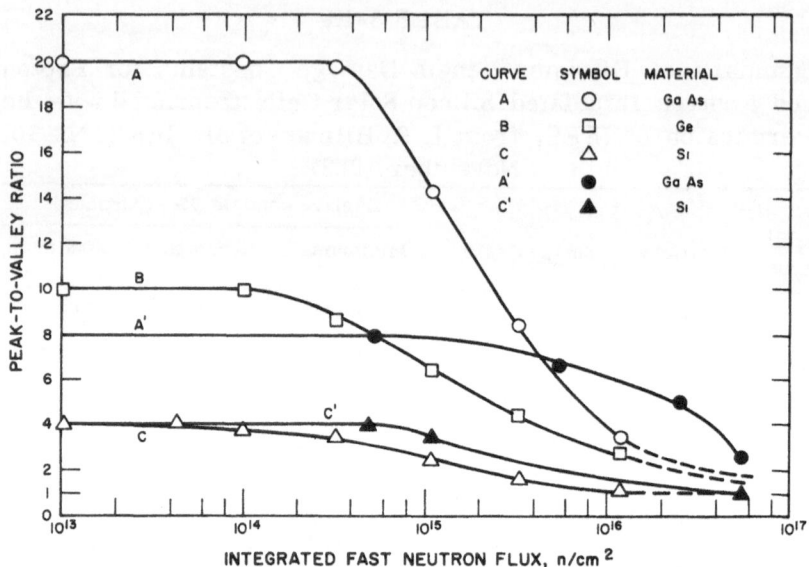

Fig. 2-14. Neutron degradation of tunnel diodes in the GTR reactor spectrum (reproduced with the permission of the Texas Instrument Semiconductor Department, from L. Taylor, Texas Instrument Report, March, 1962).

future may be evaluated by the reader in the manner outlined for the various devices above. First express the device operation in terms which include those functional characteristics known to be sensitive to radiation. Next evaluate the magnitude of the radiation-induced change to discern whether it is of significant amplitude when measured against the natural parameter of the device; for instance, a very short minority lifetime is only minutely altered by radiation damage. Finally perform radiation experimentation to adjust the theoretical constants toward a realistic value. In the majority of cases a semi-empirical equation can be developed which describes the behavior of the device with particle dose.

Ionization Effects

The ionization effects in a semiconductor material will give rise to the disturbance in carrier concentration equilibrium that was discussed earlier. For a p-type extrinsic material the situation immediately following the radiation pulse causing the ionization effects is given by:

$$n = n_i + \Delta n \exp{(-t/\tau)} \qquad (2\text{-}20)$$

where n is the instantaneous electron concentration, n_i is the thermal-equilibrium electron concentration, Δn is the ionization-induced electrons, and τ is the electron or minority carrier lifetime.

The ionization effects, which are sometimes called transient effects, are short-term, temporary disruptions of transistor operation due to radiation-generated excess carriers in the device. The effects seldom cause direct permanent damage to semiconductor devices, although for certain circuit configurations the excess current generated may cause failure of circuit functions. It can happen under conditions where a transistor is biased near breakdown. The radiation-induced current will then shift the transistor operation into its breakdown region, which could eventually lead to device burnout. Experimental data from a fast gamma–dose-rate test [17] report on such apparent failure in 2N268 transistors. The occurrence of such failure should be nonexistent in circuits specifically designed for operation in an environment where ionization effects are expected.

The ionization effects will occur when enough energy is supplied to electrons in the valence band to lift them into the conduction band at a rate which is higher than the recombination rate. The average energy that must be supplied is approximately 3 eV for germanium and 3.5 eV for silicon [18]. The energy can be applied by any form of ionizing radiation, such as charged particle, gamma or X-radiation. High-energy neutron radiation ionizes by photonuclear interactions and by imparting enough collision energy to the lattice atoms for these to become ionized.

Excess carriers are generated by ionizing radiation at a rate [19] given by

$$g = \frac{\dot{R}(t)\,\rho}{E'} \qquad (2\text{-}21)$$

where $\dot{R}(t)$ is the energy deposition per unit time (ergs/g-sec), ρ is the material density (g/cm^3), and E' is the energy required to produce an electron–hole pair (ergs), where $E' = 4.8 \times 10^{-12}$ ergs for germanium and 5.6×10^{-12} ergs for silicon. Then g is electron–hole pairs generated per unit time in a cubic centimeter of material. The term g will appear in equations to follow.

Ionization Effects in pn Diodes. The ionization-induced current in a semiconductor device has been termed the primary photocurrent, I_{pp}, given in amperes. It consists of a prompt component, I_p, being generated during the ionization period and a delayed component, I_d, which describes the decay of the induced current after the ionization source has been removed.

The Back-Biased Diode. The behavior of the back-biased diode under ionization conditions is of interest to the electronics engineer, as this diode's function in a circuit is to prevent current flow in the reverse direction. The ionization will cause a large undesired reverse current to flow during, and for a short time after, the ionization pulse.

A back-biased diode has a wide depletion region, i.e., a region on each side of the pn junction where no mobile carriers exist. It is in this region that the ionization produces the prompt component of the primary photocurrent. By utilizing the basic equation of continuity with terms for diffusion, recombination and generation included [20], the following equation is obtained for holes in the n-material and a similar equation for electrons in the p-material:

$$\frac{\partial P_n(x, t)}{\partial t} = \left[D_p \frac{\partial^2 P_n(x, t)}{\partial x^2} \right] - \left[\frac{P_n(x, t) - P_{n\text{in}}}{\tau_n} \right] + g(t) \qquad (2\text{--}22)$$

$$\underbrace{\qquad\qquad}_{\text{(diffusion)}} \qquad \underbrace{\qquad\qquad}_{\text{(recombination)}} \quad \underbrace{\qquad}_{\text{(generation)}}$$

I_p is found to be

$$I_p(t) = eAWg \qquad\qquad 0 < t \le t_p \quad (2\text{--}23)$$
$$t_p \ll \tau$$

where e is the electronic charge, A is the area of junction (in cm^2), W is the width of the depletion region (in cm), and g is the generation rate as defined in equation (2-21).

The delayed component I_d is given by

$$I_d(t) = eA \, \text{g} \, t_p \left[\frac{\sqrt{D_n} \, (e^{-t/\tau_n}) + \sqrt{D_p}(e^{-t/\tau_p})}{\sqrt{\pi t}} \right] \quad \begin{array}{l} t > t_p \\ t_p \ll \tau \end{array} \quad (2\text{-}24)$$

where e is the electronic charge, A is the area of junction, g is the generation rate, t_p is the ionization source pulse width, D_n is the electron diffusion constant, D_p is the hole diffusion constant, τ_n is the electron lifetime in p-type material, and τ_p is the hole lifetime in n-type material. I_{pp} is then given by

$$I_{pp}(t) = I_p(t) + I_d(t)$$

For diodes where τ_n and τ_p are very short with respect to t_p, which is often the case with gold-doped switching diodes, the ionization source can be considered a steady-state generator and the primary photocurrent is

$$I_{pp} = eA \text{g}(W + L_p + L_n) \quad\quad (2\text{-}25)$$

where L_p is the hole diffusion length, L_n is the electron diffusion length, and the other terms are as given above.

Forward-Biased Diode. It may be noticed that the continuity equation (2-22) does not include the electric field or drift term. This of course in the case of the back-biased diode will cause only small errors in the calculation of primary photocurrent.

J. L. Wirth and S. C. Rogers [20] in their calculations of effects on heavily forward-biased diodes included the electric field term and found that for a condition equivalent to a steady-state ionization environment and for large fields corresponding to the forward-biased diode condition, the following equation applies:

$$I_{pp} = eAW\text{g}$$

They further established that at high forward-bias voltages the resistance of the diode bulk material and the differential resistance of the junction form a self-regulating system, which tends to maintain a constant diode current in spite of the ionization-induced disturbance.

Transistors. The transient response of transistors is somewhat more complicated than that of the simple diode junction. As in the diode, the total induced primary photocurrent I_{pp} consists of two parts, namely, the prompt induced current occurring during the ionization pulse and the delayed current lasting beyond t_p , where t_p is equal to the ionization pulse width. The studies carried out in this area have all been aimed at developing a prediction technique which would yield a method for determining the I_{pp} of a particular transistor by non-irradiation means. However, due to the vast variety of transistors in terms of electrical and constructional parameters, no one universally applicable method has as yet been evolved. Of great interest to electronics engineers are the studies performed by E. A. Carr [21], who has endeavored to provide electrical parameters that may be easily measured and used to predict ionization effects response for silicon *npn* and *pnp* planar and mesa transistors. His prediction method is currently only applicable to low-power transistors whose maximum continuous collector dissipation does not exceed 0.8 W.

As with diodes, the transistor mathematical models are all based on the continuity equation with the exclusion of the electric field term. The inclusion of the electric field term in the equation was shown as necessary by J. L. Wirth and S. C. Rogers [20] to account for the drift fields in high-resistivity collector regions at high current levels. Experimental data [20] in Fig. 2-15 show the effect of the electric drift field on the delayed component of the collector primary photocurrent. It is shown how at the high exposure level [1.4 rads (Si)] instead of the normal exponential decay of the delayed component, an exponential rise is experienced until at 40 mA of induced collector current, the normal exponential decay sets in. This phenomenon is caused by an increasing drift field that is being accentuated by collector multiplication, a condition whereby minority carriers generated in the collector region under influence of drift field move into the base region. Once there the additional base current causes more collector current and yet higher electric drift fields. The low dose of 0.12 rad(Si) exposure shows the normal exponential decay at ionization pulse termination under small drift-field conditions.

There are three general conditions under which the transistor responses have been studied: (1) transistor saturated by the radiation pulse; (2) transistor not saturated and t_p, the

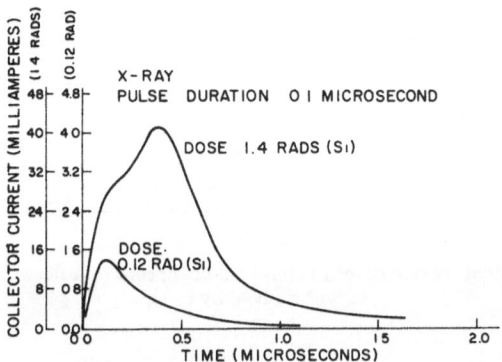

Fig. 2-15. Collector photocurrent waveforms for a 2N1051 transistor.

ionization pulse width, greater than τ, the minority carrier lifetime in the base region; and (3) transistor not saturated and t_p smaller than τ.

For the second condition the response is proportional to the ionization dose rate, whereas in the third condition the response is proportional to the dose of the ionization pulse.

For prediction of first condition response large signal transistor models are used. One of the following are normally used: (1) Ebers–Moll [22]; a saturated transistor is considered to be two back-to-back transistors in parallel. (2) Charge Control [23]; currents are shown proportional to charge stored within the transistor. (3) Linvill; the continuity equation being approximated by a difference–differential equation allowing representation by lumped elements.

The reader will now be presented with the method developed by E. A. Carr [21] which yields a good accuracy for predicting the transient response from measurement of electrical parameters. The technique has the limitations outlined above, namely, the absence of the drift field term.

The primary photocurrent (collector-base) is given by the following equation [19, 21]:

$$I_{pp} = \underbrace{\left(e \, A \, g \, \sqrt{\tau D}\right)}_{(1)}\left[\underbrace{\left(u(t) \, \mathrm{erf} \, \sqrt{\tfrac{t}{\tau}}\right)}_{(2)}-\underbrace{\left(u(t - t_p) \, \mathrm{erf} \, \sqrt{\tfrac{t - t_p}{\tau}}\right)}_{(3)}\right] \quad (2\text{-}26)$$

where the numbered terms fit the portions of the response

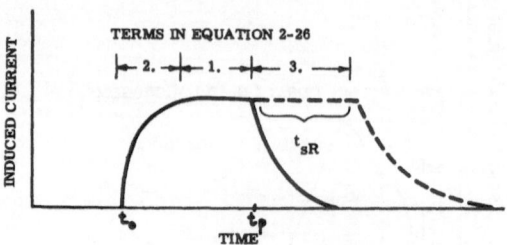

Fig. 2-16A. Typical response of primary photocurrent to a square ionization pulse of width given by t_p - t_0 .

shown in Fig. 2-16A. The terms in (1) are as defined above in the diode discussion. The other terms are defined as follows: $u(t)$ is the unit step function at $t = 0$, $u(t-t_p)$ is the unit step function at $t = t_p$ where t_p is the ionization pulse width, and erf is the error function.

Two terms, τ (minority carrier lifetime in collector) and A (junction area—coll. to base), must be uniquely determined before the equation can be solved. The value τ may be electrically determined by the determination of t_s, the collector storage time (μ sec); t_s is measured electrically using the circuit shown in Fig. 2-16B, and then inserted in the following equation:

$$ \tau = \frac{t_s}{\left\{ \mathrm{erf}^{-1} \left[1 -(I_{cs}/\ h_{FE}\ I_B) \right] \right\}^2 } \qquad (2\text{-}27) $$

where I_{cs} is the collector saturation current. A, the collector junction area, may be available from the device manufacturer;

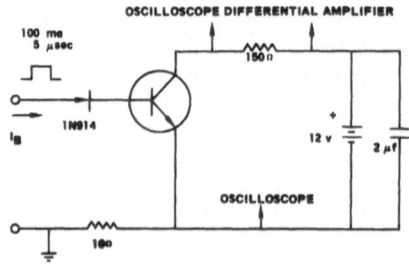

Fig. 2-16B. Circuit for measuring storage time t_s.

if it is not available it may be determined by inserting t_s in the following empirical equation [21]:

$$\ln\left[(h_{FE} + 1)\,A\right] = 0.67\,\ln t_s - 2.54 \qquad (2\text{-}28)$$

where A is measured in square centimeters and t_s in microseconds; h_{FE} is the pulsed value at 20 mA I_c. With A and τ determined, equation (2-26) will yield the magnitude and shape of the primary photocurrent for a given ionization pulse in silicon planar and mesa transistors.

As the ionization dose rate increases and the transistor begins to saturate the response prediction must be modified to include the radiation storage time t_{sR}, measured in microseconds. Figure 2-16A shows that t_{sR} is defined as the extent of time that a transistor remains in saturation after termination of the radiation pulse. E. A. Carr does not include the drift currents in his continuity equation calculations, which as shown by Wirth and Rogers in certain instances will lead to inaccurate predictions at times immediately after the termination of the ionization pulse. Nevertheless, E. A. Carr has for the conditions he has chosen been able to produce good correlation between theoretical and experimental determination of t_{sR} for silicon planar and mesa transistors. t_{sR} is as may be expected related to t_s, the electrical collector storage time, and E. A. Carr has experimentally shown that a correlation exists [21]. He reports on a test involving 16 silicon planar and mesa transistors (open base configuration) subjected to a $0.2\,\mu$sec ionization pulse. From this data an empirical equation is evolved giving t_{sR} in terms of t_s and ionization dose rate:

$$t_{sR} = 0.282 t_s \ln \dot{\gamma} - 5.11 t_s \qquad (2\text{-}29)$$

$$(I_{cs} = 80 \text{ mA})$$

where $\dot{\gamma}$ is the ionization dose rate [rads (Si)/sec].

For the cases where the drift currents are judged by the reader to be important the drift field term must be included in the continuity equation. The complete derivation is given thoroughly by Wirth and Rogers [20].* Wirth and Rogers point

*This paper may be obtained from the Nuclear Science Group of the Institute of Electrical and Electronics Engineers, New York [20].

to the important fact that by including the electric drift field term the continuity equation has become nonlinear. Primary photocurrents are therefore not dependent on the mode of device operation and do not necessarily scale linearly with dose and dose rate. Experimentally determined primary photocurrents for a reverse biased collector to base junction should therefore be used only with great caution.

The magnitude of the errors introduced in the I_{pp} prediction by either neglecting to include the electric drift field term or by using experimentally determined I_{pp} depends on the following:

1. The design of the transistor. High-resistivity collector transistors cause greater error than do low collector resistivity transistors.
2. The circuit design. If the circuit causes the transistor to saturate at low collector currents the drift field may not be significant.

The Secondary Photocurrent, I_{sp}. The secondary photocurrent I_{sp} is that portion of the primary photocurrent which crosses the emitter-to-base junction and in the process is amplified by β. The total ionization-produced current which flows in the transistor collector external circuit is then the sum of I_{pp} and I_{sp}:

$$I_{C_{ion}} = I_{pp} + I_{sp}$$

where the subscript *ion* represents ionization. I_{sp} is given by

$$I_{sp} = I_{pp}\beta(1 - f) \qquad (2-30)$$

The fraction f is determined by internal and circuit impedances in the base and emitter circuits and is given by

$$f = \frac{1}{1 + R_B/\beta RE} \qquad (2-31)$$

The MOS Transistor . The ionization effects in the MOS transistor have been shown [24] to consist of two components, drain-substrate diode photocurrent and secondary drain photocurrent. From these observations a small-signal equivalent circuit has been developed as shown in Fig. 2-17.

The photocurrent in the drain substrate diode is swept

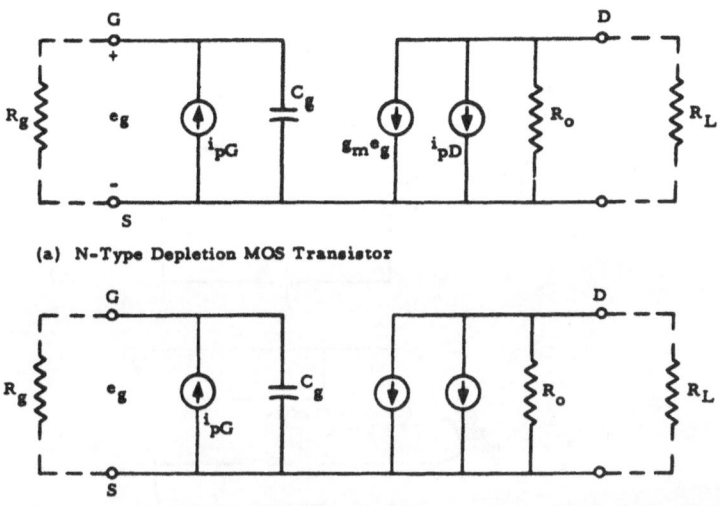

(a) N-Type Depletion MOS Transistor

(b) P-Type Enhancement MOS Transistor

Fig. 2-17.   MOS transistor.   Radiation-induced small-signal equivalent circuit.

across the junction resulting in a transient drain current. The secondary drain photocurrent is due to carriers generated in the channel. These carriers move under the influence of the existing electric field. A resultant decrease in the gate voltage will occur as the channel region charge is changed. In the equivalent circuit, i_{pD} is the drain substrate I_{pp}, I_{pG} is the gate displacement I_{pp}, and C_g is the gate capacitance.

Surface Effects

The surface effect in transistors manifests itself as an increase in reverse collector current and becomes more pronounced as the reverse bias of the collector junction is increased. The effect also causes large degradations in current gain of transistors. The effect is primarily dependent on the total ionizing dose rather than the dose rate. It is an effect strictly associated with ionizing radiation.

As the name implies, this is an effect which interacts with device surfaces. As described earlier the surface can be viewed as an extended defect in the crystal structure and its interactions with interface materials such as gases and foreign material atoms are not predictable to any good accuracy. The

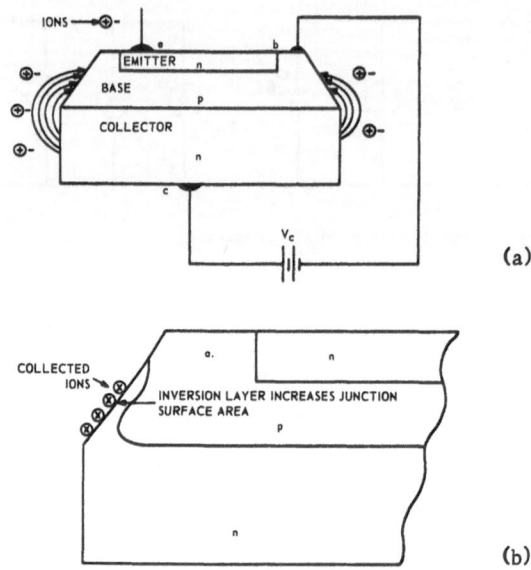

(a)

(b)

Fig. 2-18. The surface effect in the mesa transistor.

surface effect phenomenon was described in early 1963 by
D. S. Peck et al. [25], who at that time was concerned with the
influence of these effects in Telstar satellite electronic circuits.
The work by Peck et al. was initiated in 1961. Investigations
into the phenomenon were, however, reported as early as 1956
by H. L. Steele [26]. Not surprisingly the surface phenomenon
has also been plaguing nuclear physics instrumentation personnel.

The surface phenomenon, as mentioned, is not accurately
predictable and no analytical mathematical model is therefore
available. However, a theoretical model was created by Peck
et al. and has been described by R. R. Blair [27]. The model
is shown in Fig. 2-18. The exposure of the gas in the transistor can to ionizing radiation results in the production of
positive ions and electrons, which in turn are acted upon by
fringing electric fields in the vicinity of the transistor
junctions and attach to surfaces of attracting polarity. The
can-to-device electric field will also help or hinder this

process. As seen in the lower part of the figure, the collection of ions on the surface of the base region will cause an inversion layer to be formed which contains the ionization-produced electrons. This inversion layer or channel increases the reverse collector current.

That the surface effect is dose-dependent rather than dose-rate-dependent is demonstrated by Fig. 2-19. It is seen here how a change in dose rate from 8.5×10^5 rads/hr to 1.4×10^5 rads/hr does not disturb the linear relationship between integrated dose and increasing reverse collector current, I_{CBO}. Figure 2-20 shows the dependence of the surface-effect magnitude upon the content of the transistor can, and Fig. 2-21 its dependence on bias.

It was found early that devices which had a surface of some structural integrity from a macroscopic viewpoint, such as those devices with a silicon oxide coating, were less affected by surface effects. Figure 2-22 shows the comparative response of two devices, one a gas-encapsulated ordinary device

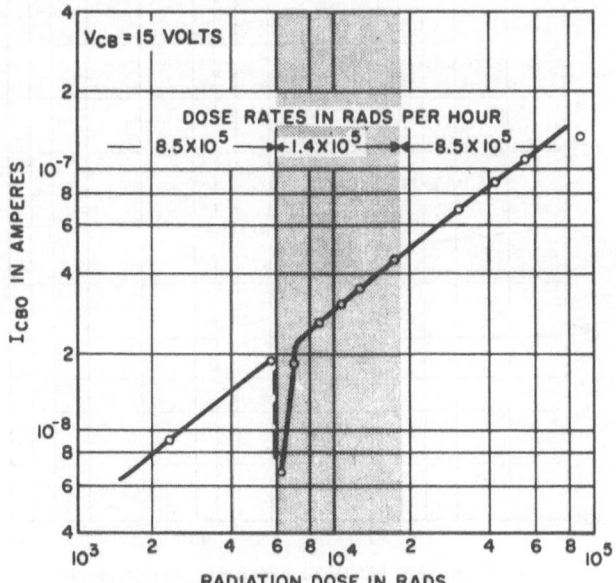

Fig. 2-19. I_{CBO} vs. radiation dose.

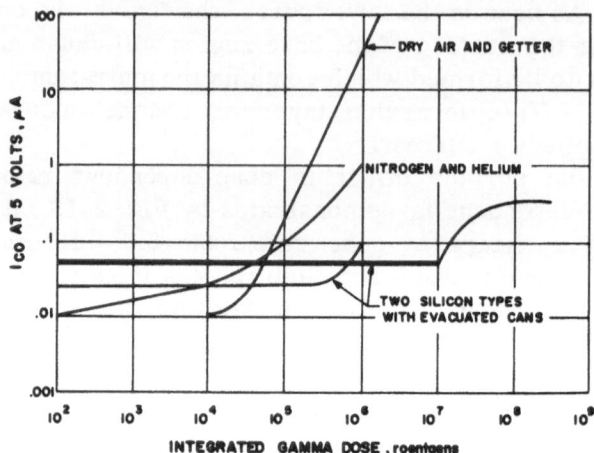

Fig. 2-20. Transistor surface-effects dependence on container gas content (reproduced with the permission of IEEE, from R. R. Blair, IEEE, NS-10, November, 1963).

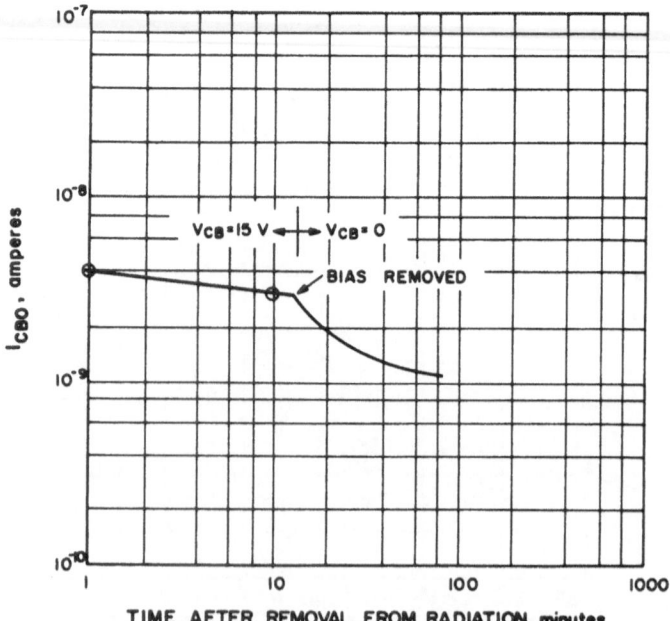

Fig. 2-21. Ionization transistor surface effects (with bias parameter) (reproduced with the permission of IEEE, from R. R. Blair, IEEE, NS-10, November, 1963).

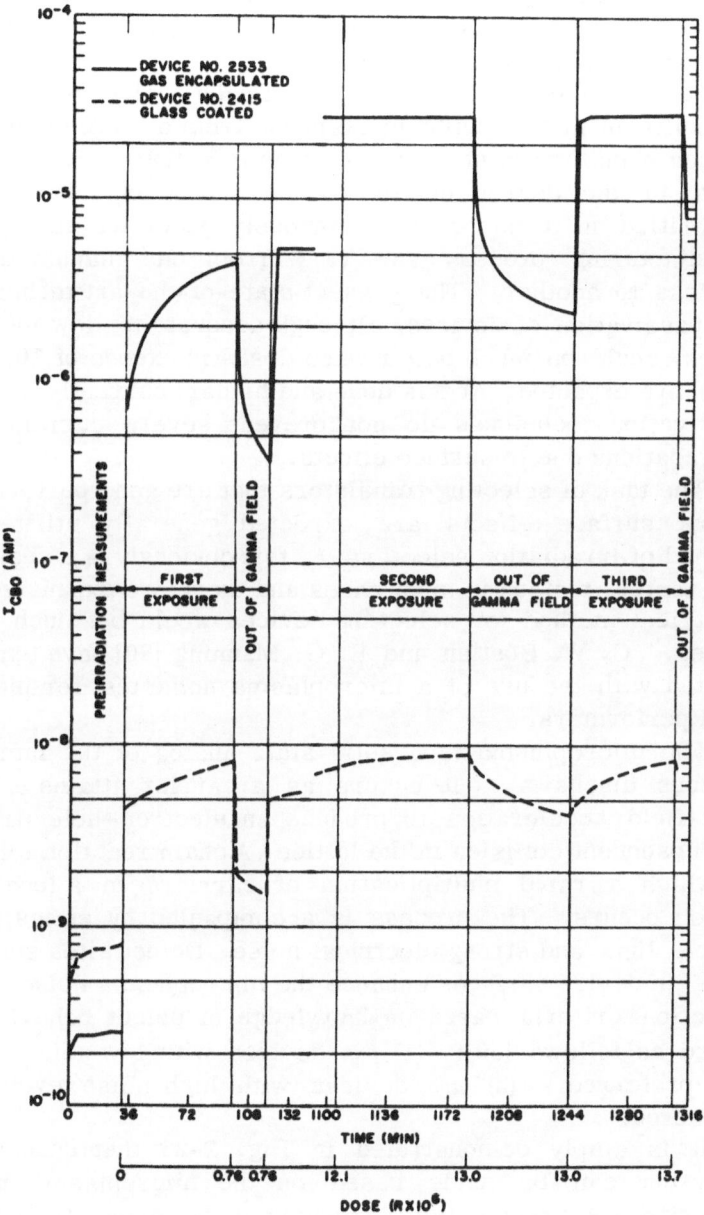

Fig. 2-22. The reverse collector-base current with emitter open I_{CBO} vs. integrated gamma dose.

and the other a glass-coated device [28]. In devices that have protected surfaces, the change in I_{CBO} is much less than for an unprotected surface device.

Experiments performed by J. C. Peden et al. confirmed the difficulty in getting uniform response from a large number of silicon planar transistors of the same type [29]. Large variations in the degree of passivation and amount of surface impurities in planar devices obviously occur within a given manufacturing process and vary from one manufacturing process to another. The present state-of-the-art techniques for passivation of devices, although adequate for low ionizing doses, could be much better when doses in excess of 10^4 rads (air) are expected. At this dose and higher, currently utilized passivation techniques do not prevent severe current gain degradations due to surface effects.

The task of selecting transistors that are good performers where surface effects are expected generally utilize the method of irradiation selection. As this obviously is a destructive testing method in most cases and therefore expensive, an electrical method for selecting devices would be much preferred. C. W. Bostian and E. G. Manning [30] have experimented with the use of a microplasma noise test for picking good performers.

The microplasma is a solid-state analog of the familiar gaseous discharge. It occurs as a carrier attains enough drift-field acceleration to produce an electron–hole pair at its subsequent collision in the lattice. A chain reaction follows in which a rapid multiplication of carriers in a localized region occurs. The process is accompanied by emission of visible light and strong electrical noise. Defect sites such as those in device surfaces enhance the microplasma noise. The selection criteria based on knowledge of defect behavior is stated as follows [30]: (1) pnp devices with low noise levels are preferred. (2) npn devices with high noise levels are preferred.

It is amply demonstrated in Fig. 2-23 that successful selection can be made based on the microplasma noise criterion. A preselection screening is performed to ensure, first, that the transistor is mechanically perfect, as mechanical defects will not be detected by the noise test and could never-

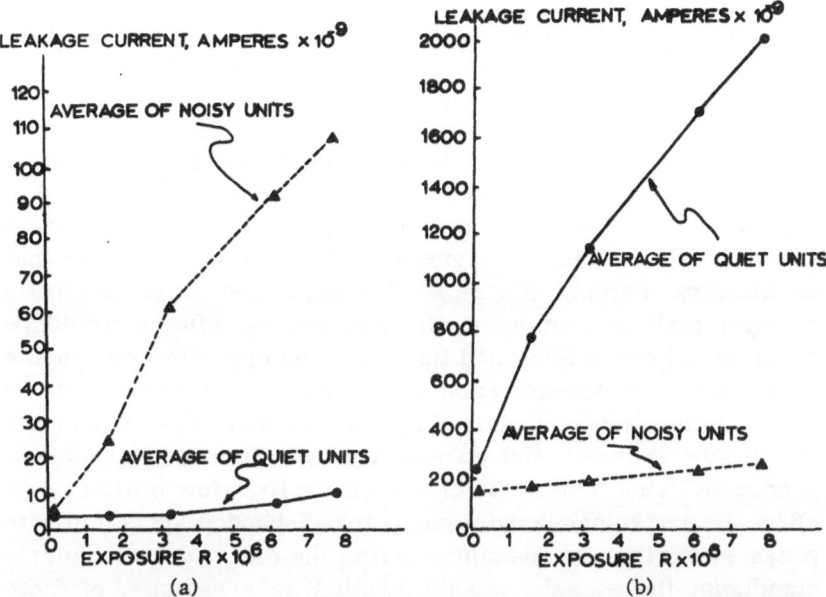

Fig. 2-23. Microplasma noise selection (reproduced with the permission of IEEE, from C. W. Bostian and E. G. Manning, IEEE, NS-12, February, 1965). (a) Average leakage-current behavior of quiet and noisy 2N2411 silicon *pnp* planar transistors during irradiation. (b) Average leakage-current behavior of quiet and noisy 2N298 silicon precision-alloy *npn* transistors during irradiation.

theless make the transistor degrade rapidly. Secondly, those transistors with reverse collector currents 50 to 100 times the average for their type are also rejected.

The impulse noise produced by microplasma discharge is a function of the reverse bias current. The noise test should therefore be performed over a range of I_{CBO}.

OTHER RADIATION EFFECTS

The Electromagnetic Pulse Effect *

The electromagnetic pulse which is produced at the time of a nuclear detonation is of considerable interest to the electronic instrumentation engineer concerned with the design of space vehicle equipment. Ordinary detonations of chemical nature produce similar signals, but not as large as those accompanying a nuclear explosion.

*Taken in part from S. Glasstone, U.S. Government Printing Office [32].

Two mechanisms are believed to create the total electro-magnetic pulse. The first is often called the Compton-electron model and is created as the initial gamma radiation collides with electrons in the atoms and molecules of the surrounding air. These Compton electrons usually move rapidly away from the center of the burst. Provided there is some kind of asymmetry, this electron motion is apparently one of the main sources of the electromagnetic pulse. It is important that asymmetry exists, since if the explosion were perfectly symmetrical in a uniform atmosphere, the effects would be equal in all directions, and therefore the opposite components would then compensate each other exactly and there would be no electromagnetic signal. In practical cases the asymmetry is always ensured and an electromagnetic signal is always produced. The signals are emitted in the first few milliseconds after the burst and have a very broad-banded spectrum with peaks at various frequencies. From the electronic designer's standpoint the signals can be handled as a bad case of radio noise and ordinary precautions against this type of noise would make the designer's system fairly insensitive to this part of the electromagnetic pulse.

Another type of signal is created by the "field-displace-ment" mechanism. It is generated as the highly ionized vapor, or plasma, as the center of the burst expands, and tends to exclude the magnetic field of the earth. A hydromagnetic wave is thereby set up which moves away from the immediate vicinity of the burst. Since this is a slowly moving magnetic field it has great penetration power, and the electronics engineer protecting against this effect must utilize high-permeability shielding to reduce its magnitude to levels that his equipment can tolerate. Cabling must be protected and standard techniques for reduction of induced voltages must be utilized.

Gamma and Neutron Heating*

When dealing with design problems involving nonshielded or at least very lightly shielded steady-state nuclear reactors —

*Taken in part from T.R. Strayhorn, Space/Aeronautics, 39 (3):111 (March 1963).

such as would be the case when building electronic equipment for operation close to nuclear propulsion systems — the phenomenon known as gamma and neutron heating becomes important. This parameter must be considered in addition to the other radiation effects which are produced by the neutron and gamma spectrum emitted by the nuclear reactor. The magnitude of the heating varies as a function of several factors, among which are the distance from the reactor, the reactor operating time, the reactor power, and the reactor gamma-neutron spectrum.

Radiation heating by neutrons is the result of collisions (elastic and inelastic) between neutrons and the nuclei of a material. This is the predominant interaction. Other interactions include (1) neutron–proton, (2) neutron–gamma, (3) neutron–alpha, and (4) neutron–beta reactions. The type of reaction occurring depends upon the material in which the reaction takes place. The reaction first listed, neutron–proton, is important in materials containing gadolinium, the third in materials containing boron and lithium. Neutrons as a rule are most important in the heating of organic materials of high hydrogen content.

The heat deposition rates of three typical materials are given in Table 2-IV. These figures are based on an RBE equal to 1 for gamma radiation and equal to 10 for neutrons. Also, an unshielded reactor is assumed, which in its spectrum

TABLE 2-IV

Heat Deposition Rate for Polyethylene, Aluminum, and Iron

Material	Density (g/cm^3)	Heat deposition rate $\left(\dfrac{\text{Btu/in.}^3\text{-sec}}{\text{rem/hr}}\right)$ *	
		Neutrons†	Gamma radiation†
Polyethylene	1.0	9.18^{-12}	4.06^{-11}
Aluminum	2.7	2.98^{-13}	1.02^{-10}
Iron	7.8	1.43^{-13}	3.09^{-10}

*Rem/hr = 82.16 rads (C)/hr.
†Exponent is power of ten, i.e., 9.18×10^{-12}.

has a high amount of thermal neutrons. A smaller reactor such as a SNAP type probably would be shielded to some extent.

The temperatures reached by the materials of nuclear-powered vehicles depend on the total heat input and the amount of cooling; these, in turn, depend on the particular vehicle and its mission. Since the only cooling available in the vacuum of space is by radiation, it is difficult to make efficient cooling systems within the weight and volume restrictions that exist. Furthermore, as materials are heated by the radiation, a point is reached where nuclear heat input equals the thermal radiation output. A typical situation is shown for a 10,000 mW reactor, and it can be seen that only by increasing the distance from the reactor can the temperature be kept down.

In most of the earlier nuclear missions that are planned, the reactor will operate only for a short period. Figures 2-24 through 2-27 show the time needed to achieve the equilibrium temperature vs. distance from the reactor, and the reactor power. To achieve equilibrium temperature takes an infinitely long time, and only the time needed to reach 0.95 of T'_{eq}, the

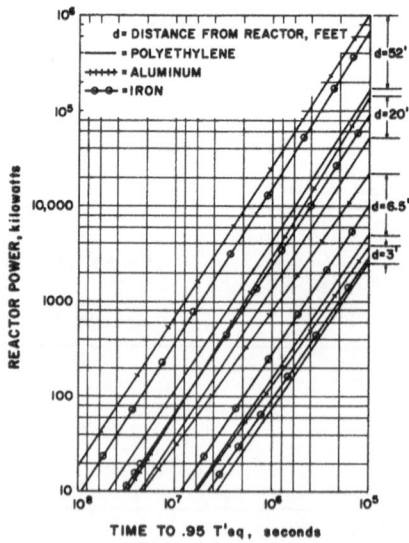

Fig. 2-24. Reactor-induced heating, time to approach equilibrium temperature.

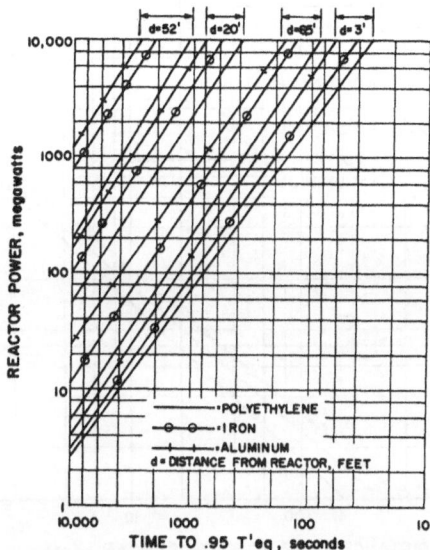

Fig. 2-25. Reactor-induced heating, time to approach equilibrium temperature.

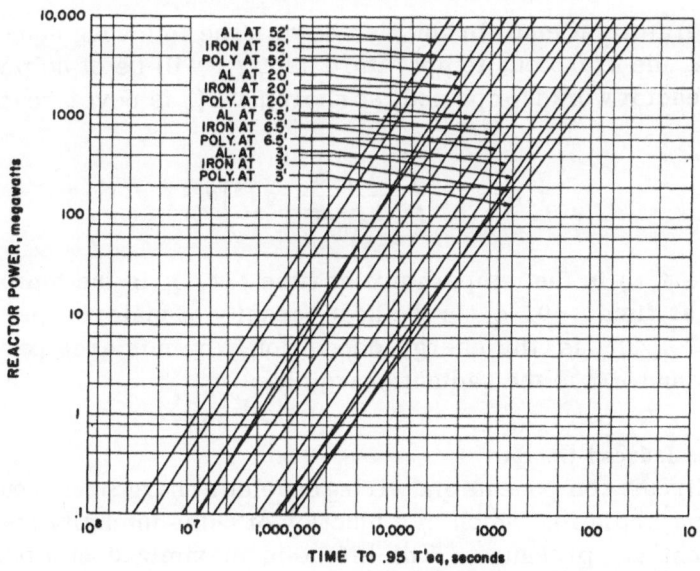

Fig. 2-26. Reactor-induced heating, time to approach equilibrium temperature.

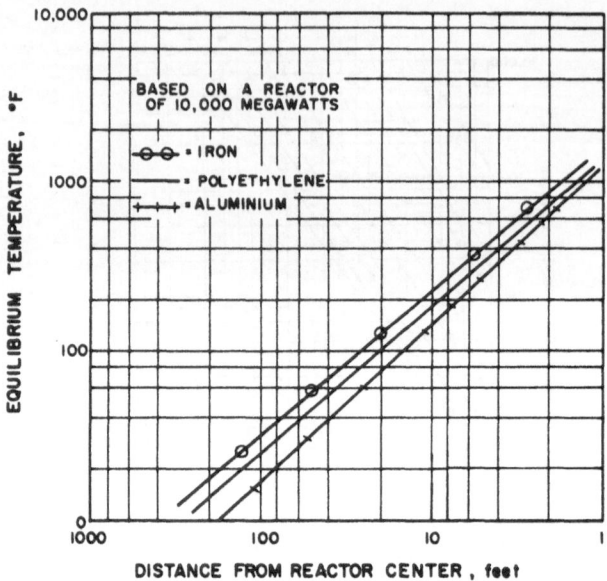

Fig. 2-27. Reactor-induced heating, equilibrium temperature.

temperature at equilibrium, is shown. The following equation, where the transient temperature is $T_t(t)$, will be of help when the reactor works for periods so short that T'_{eq} is never reached:

$$T_t(t) = T_0 + \int_0^t \frac{1}{c_p}\left(\frac{dq}{dt} - \sigma T^4\right) dt$$

where $T_t(t)$ is the temperature at time $t = t$, T_0 is the temperature at time $t = 0$, c_p is the heat capacity of material per unit area, dq/dt is the energy deposition per unit area per unit time, and σT^4 is the radiant energy loss.

Gamma-Induced Charge

Circuit components and wiring develop a charge under pulsed gamma radiation which is a function of pulse intensity and the ambient air pressure. The effect is maximized at a partial vacuum of around 10 mm Hg and is virtually eliminated at a vacuum of 10^{-3} mm Hg. The voltage corresponding to this charge has a prompt component, which coincides with the gamma pulse width, and a decay component which persists for several times this width.

SUMMARY OF MATERIAL EFFECTS

Radiation Effects*

The radiation effects on vacuum tubes include (1) slight decreases in plate current, (2) fractures and failures of glass envelopes, and (3) erratic grid and filament currents.

The radiation effects on gas tubes include (1) slight increases in plate voltages, (2) some increases in plate current, and (3) some fractures of glass envelopes.

Radiation effects on microwave and light-sensitive tubes include (1) some separation of anode caps from the glass, (2) considerable fluctuations in the dark current, and (3) considerable discoloring of the glass envelopes, causing loss of sensitivity.

There is very little radiation effect on glass dielectric capacitors.

In mica dielectric capacitors, there is little permanent effect. Transient increases in capacitance occur.

There is no appreciable radiation effect on ceramic dielectric capacitors. Capacitance generally stays within tolerance, although a fast neutron flux does affect insulation resistance.

In the study of radiation effects on paper dielectric capacitors, it was found that (1) capacitors exposed directly to the pile suffer gas evolution, (2) capacitance tends to decrease rather than increase when tolerance limits are exceeded, and (3) insulation resistance degrades to the extent that some test units shorted or ruptured.

Radiation effects on plastic dielectric capacitors are as follows: (1) Some units suffer drastic distension and short-

*The effects information presented here has been abstracted from a report entitled "The Effect of Nuclear Radiation on Electronic Components" issued by the Radiation Effects Information Center (REIC) of the Battelle Memorial Institute, Columbus, Ohio, in June 1961. Consider for information only and obtain updated data from REIC.

circuiting; other units experience no AC loss and little change in capacitance.   (2) The dissipation factor varies slightly. (3) Impregnated oil polymerizes under neutron flux.

Radiation effects on electrolytic capacitors are as follows: (1) Capacitance values vary and tend to increase. Most variations stabilize within three days after dosage is removed; however, some are permanent.   (2) Dissipation factors tend to decrease as much as 24%.

Insulators have been tested for the response of both electrical and mechanical properties to neutron flux dosage; typically it may be said that (1) vinyl plastic suffers severe discoloration, (2) polypropylene and silicone rubber become brittle and lose tensile strength, (3) mean dielectric strength of mica is degraded as much as 60%, and (4) water absorption of Melmac is degraded 50%.   Asbestos suffers no net change in insulation resistance if it is protected from thermal neutrons.

The study of radiation effects on transformers reveals that electrical performance is essentially unaffected, but there is general rupture of hermetically sealed cases. Synchrotransformers are affected more than filament, audio, and IF transformers.

Radiation effects on coaxial cables include (1) potential drops along conductor, (2) saturation of insulation with respect to voltage applied, and (3) coaxial shield breakdown.

Special Circuit Design Considerations for the Nuclear Environment

During a recent conference devoted to discussion of problems in solid-state circuit design, several important points were brought out. There was general agreement that the first effects of dosage increase are felt by transistors. They are more sensitive by as much as three orders of magnitude than other circuit elements.

Uniformity in nuclear radiation susceptibility is very poor. Workers in the field hope that a reduction in semiconductor susceptibility will improve the level of uniformity. Among measures that are being considered is gold doping. This would reduce minority carrier lifetime of some portions of devices which are not important to the control of performance in present environments.

TABLE 2-V
Results of Commercial Testing Procedures

Device	nvt Values
Very high-speed planar (2N709)	10^{15} *
Power transistor	$2-5 \times 10^{13}$ †
Planar epitaxial (2N919) ‡	5×10^{15} †
Planar device (perhaps 2N708 type) ‡	10^{14} †
Planar diode ‡	2×10^{17} †
Silicon diode (200−400 V type)	2×10^{15} †

* $\beta = 10$ for $I_e = 1$ mA.
†No permanent effect.
‡Device was gold-doped to control base and collector lifetimes in order to obtain a minimum storage time.

Certain commercial testing procedures use Van de Graaff-generated 2 MeV electrons. The data are scaled on the basis that 2 MeV electrons are about 25 times less effective than neutrons in degrading devices. Equipment is tested in continuous and pulsed reactors. The data which have been obtained are presented in Table 2-V. The criterion for diode acceptance is a current of 100 ± 30 mA at a potential drop of 1 V. The criterion for transistors is that the β be greater than 10 with no permanent effect.

Testing has indicated that we do not as yet have optimized transistors for maximum exposure to radiation. It seems that the best device for high radiation tolerance is an epitaxial planar of very high speed in which the collector saturation resistance is compromised. Of course, the best devices are obtained only when each is designed for its own specific use in a nuclear environment. The present integrated circuits do not use epitaxial devices; the complete circuit is grown in an epitaxial layer.

The problem of reducing surface effects due to ionization of gas in a transistor package becomes troublesome when the dosage exceeds two orders of magnitude less than that which produces permanent damage. Evacuation would seem to be an

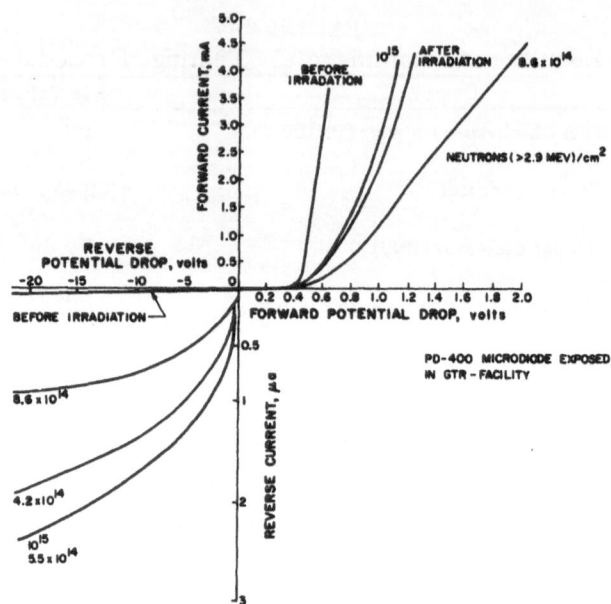

Fig. 2-28. Pacific semiconductors, microdiode PD-400 degradation in the GTR environment. Reported levels are > 2.9 MeV.

easy solution, but transistor efficiency is reduced when no gas is present in the package. It has been agreed that passivated surface devices are less susceptible to surface charge effect than are unpassivated devices.

Resistance to interference produced by pulsed radiation is another serious problem for circuit designers. An apparent compensation is a circuit which employs a diode in reverse across the base-to-emitter diode of the transistor to compensate for the leakage produced by the pulse. The problem here is to find the correct diode for each transistor application.

Figures 2-28 to 2-32 and Table 2-VI are included as reference material for the reader. It should be used with caution because thorough reporting of all experimental conditions is beyond the scope of the book. The data are worthwhile in that they indicate the trends the reader should be watching for in his own experimental data.

RADIATION EFFECTS ON MAN

A handbook on radiation effects would be incomplete if it

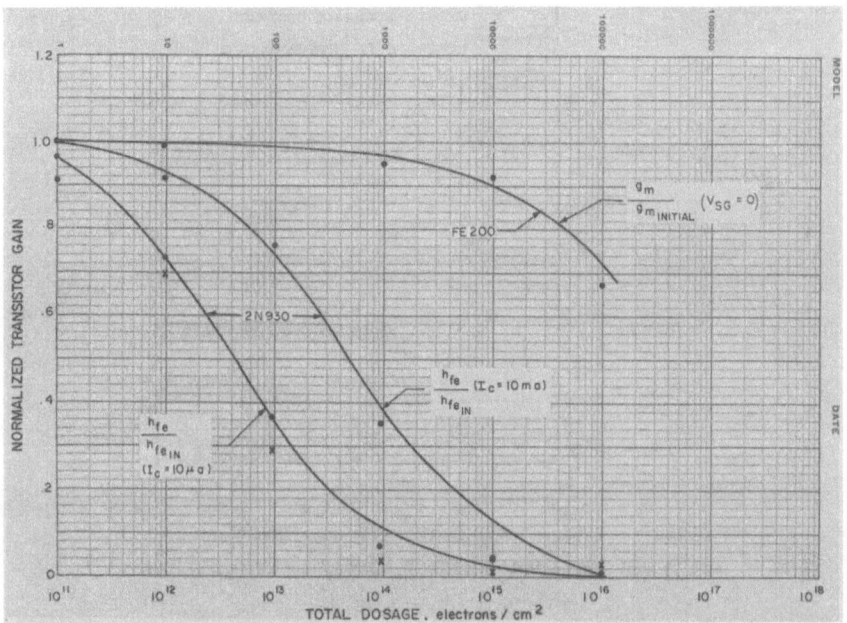

Fig. 2-29. Degradation of field effect transistor FE200 and transistor 2N930 in response to 1 MeV electrons (reproduced with the permission of Amelco Semiconductor).

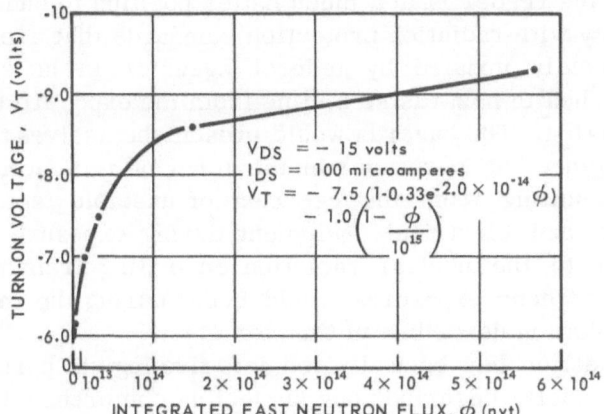

Fig. 2-30. Turn-on voltage vs. neutron flux for a p-type enhancement MOS transistor (reproduced with the permission of IEEE, from G.C. Messenger et al., IEEE, Nuclear Science Group Meeting, July, 1965).

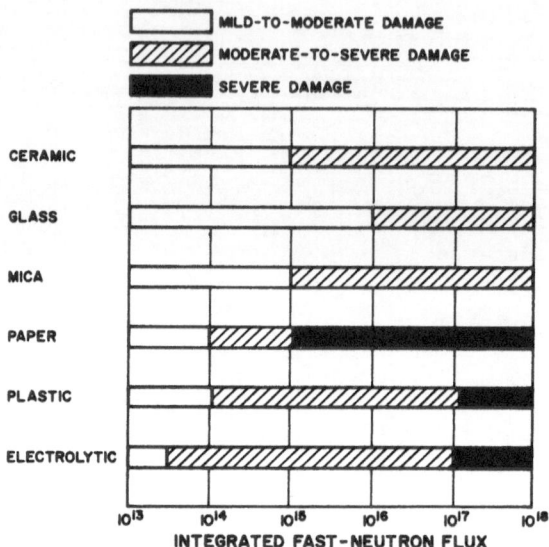

Fig. 2-31. Relative radiation sensitivity of capacitors.

did not contain a section devoted to the effects of nuclear radiation on the human body, since exposure to radiation is always destructive to living cells no matter how small the absorbed dose. With a thorough knowledge of these effects, however, tne reader is in a much better position to understand and comply with radiation protection standards that are issued and frequently updated by federal agencies in an effort to minimize health hazards stemming from the exposure to ionizing radiation. The hazards would present themselves to electronic engineering personnel in the form of radioactive materials resulting from the creation of unstable isotopes in electronic and electrical equipment during exposure of this equipment to the nuclear radiation emanating from nuclear reactors. These exposures would occur during the radiation experimentation described in Chapter 4.

The section has been divided into five logical parts, each of which briefly describes one subject as comprehensively as is possible, although radiation effects on man are far from well understood due to the fortunately limited amount of research material available to scientists. The five divisions are (1) radiation exposure and biological effects, (2) comparative risks from radiation sources, (3) somatic effects, (4) hereditary effects, and (5) radiation protection standards. These divisions are summarized as follows:

1. Radiation Exposure and Biological Effects. The main questions discussed here are how does radiation interact with living organisms and how is this expressed quantitatively?

2. Comparative Risks from Radiation Sources. As the reader has found in Chapter 1, there are many types of radiation, each influencing living organisms in ultimately the same way, but some radiation types damage more severely than others for the same absorbed dose. It is the purpose of this section to provide a quick comparison among the different types.

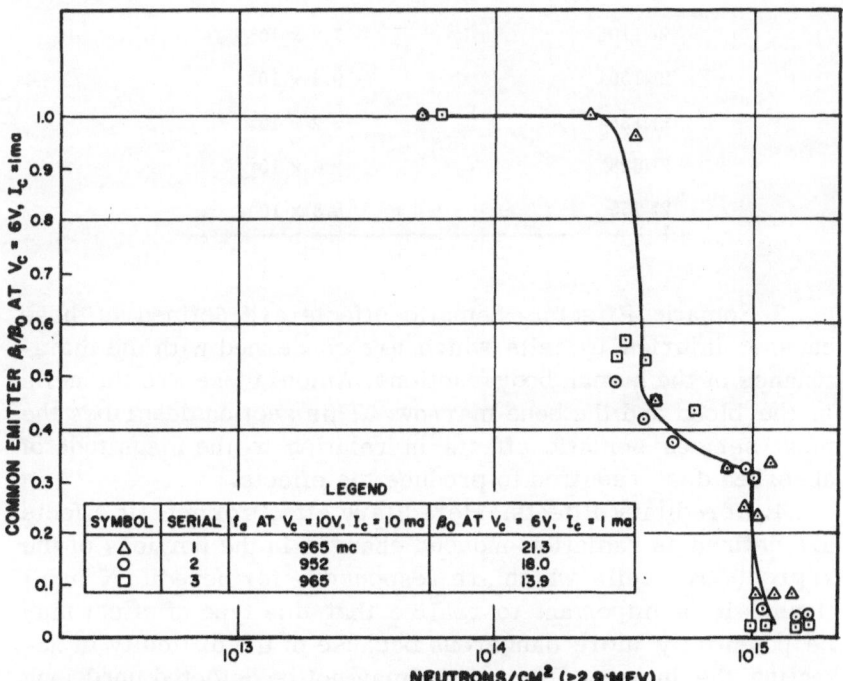

Fig. 2-32. Degradation of a *npn* germanium mesa transistor 2N797 in a GTR (Ground Test Reactor) spectrum [12]. Neutrons (> 2.9 MeV)/cm². Note the three degradation regions described in the discussion on displacement effects, Fig. 2-6.

TABLE 2-VI

Typical K_r Values

Transistor	$K_r \left(\dfrac{\text{neutrons-sec}}{\text{cm}^2} \right)$
2N726	2.5×10^6
2N743	2.3×10^6
2N1406	9.5×10^6
2N700	1.4×10^8
2N797	1.6×10^6
2N706	9.7×10^6
2N753	3.4×10^6
2N834	1.1×10^6
2N1141	1.80×10^6
2N1195	1.3×10^6
2N1561	9.3×10^4
2N1385	9.2×10^5
2N828	2.6×10^5
2N769	8.8×10^5

3. Somatic Effects.   Somatic effects are defined as those causing injuries to cells which are concerned with the maintenance of the human body functions. Among those are the cells in the blood and the bone marrow. This section describes the most serious somatic effects in relation to the magnitude of absorbed dose required to produce the effects.

4. Hereditary Effects. Hereditary effects or genetic effects are defined as radiation-induced changes in the portions of the reproductive cells which are responsible for hereditary functions.   It is important to realize that this type of effect may be potentially more dangerous because of the difficulty in detecting the injury.   The injury may not be detected until long after the radiation accident, and will then often be erroneously associated with other causes.

5. Radiation Protection Standards. The maximum permissible absorbed radiation doses are described and explained. The reader will find reference to the various radiation monitoring equipment available, as well as guidance for protection from dangerous radiation sources.

Radiation Exposure and Biological Effects

The human body as a living organism is composed of tissues which are made up of cells of various types. The destruction of the cells signifies an injury. Nuclear and high-energy atomic radiation can produce the injury, when the radiation is absorbed by the living cell. It appears that the harmful effects on the cells are caused by the energy released as ionization when the radiation is absorbed. The injury can take various forms, but can generally be classified as cell death, impaired cell functions, difficulties in cell division, and changes in gene structure.

The assessment of radiation in terms of its capability to produce harmful effects in human organs is broken down into two tasks:

1. Assessment of the efficiency with which the radiation produces ionization.
2. Calculation of the total dose absorbed by the tissue.

The dose rate is also of interest, as it has been found that for equivalent doses greater harmful effects were observed when the dose was delivered in a short time. As described above in the analysis of the types of radiation, there are generally two types of ionizing radiation. The radiation is either directly ionizing or indirectly ionizing. Directly ionizing radiation consists of charged particles, such as beta particles (electrons and positrons of nuclear origin), electrons, alpha particles (helium nuclei, which have two positive charges), and heavier ions (atoms stripped of some of their electrons). Indirectly ionizing radiation consists of high-energy electromagnetic radiation (X- and gamma rays) and neutrons. These types of radiation when interacting with materials produce ions in the exposed materials, which subsequently cause ionization. The types of radiation particularly important from the radiation safety standpoint are those emitted by radioactive materials, i.e., alpha particles, beta particles, and gamma and X-radiation. These will be described along with the neutron, which may also be encountered during nuclear reactor experimentation.

Alpha particles are doubly charged, and are as a consequence extremely ionizing. They have therefore very limited penetration power, and are easy to protect against. They occur mainly as secondary radiation from slow neutron captures and as emissions from high-density radioisotopes such as radium-226. The 4.77 MeV alpha particle from radium-226 has a range in air (15°C and 760 mm pressure) of only 3 cm, and will not penetrate the protective layer of the skin.

Beta particles are electrons and positrons of nuclear origin. For a given energy, beta particles have a larger range than alpha particles. The beta particles from radioisotopes are emitted with a continuous energy spectrum. Tables showing decay by beta emission list the maximum beta energy emitted, and for practical purposes a mean energy of $\frac{1}{3}$ the maximum is often used. Below an energy of approximately 1 MeV the beta particles lose energy directly by ionization. As the energy increases beyond 1 MeV, the loss of energy is predominantly more and more by the emission of bremsstrahlung (braking radiation), in which electromagnetic (X-) radiation is produced. The bremsstrahlung ultimately spends its energy by indirect ionization. Beta emission is an important emission from a variety of radioisotopes of importance to electronics engineers.

Gamma and X-radiation are electromagnetic radiations. Gamma radiation has a nuclear origin whereas X-radiation is produced in the electronic bands of the atom. It is of course quite possible to have X- and gamma radiation of the same energy, and the effects caused by either are identical. The method by which ionization is produced was described above in the analysis of types of radiation. Gamma radiation is the other very important emission from radioisotopes, and because of its great penetration power it is potentially the most dangerous from a biological standpoint.

The radiation unit most applicable to the subject of radiation on living cells is the rad. As described in Chapter 1, it

TABLE 2-VII

RBE Values for Types of Radiation

Radiation	RBE
X-, gamma, and beta radiation	1
Thermal neutrons	2.5
Fast neutrons	10
Protons	10
Alpha radiation	10
Recoil nuclei	20

is a unit of absorbed dose equal to 100 ergs/g of material. The radiation absorbed by a living organism produces a change. The biological effect on the other hand may also depend upon the distribution of the energy released along the track of the ionizing particle. To take this into account the unit "RBE" (relative biological effectiveness) was introduced. It is defined as the following ratio:

$$RBE = \frac{\text{Absorbed dose (250 keV X-rays)}}{\text{Absorbed dose (other radiation)}}$$

where the absorbed dose is that dose which for both cases produces the same biological effect.

The biological effect is therefore not only a function of the absorbed dose in rads but also of the RBE for the particular type of radiation. This leads then to the unit most often used in radiological protection work [31], the rem,

$$rem = (rads)(RBE)$$

Rem is the acronym for "roentgen equivalent man." Table 2-VII gives RBE values for different types of radiation. These numbers are only recommended values; for instance, in the case of fast neutrons it is argued that the number is rather high, and that a much lower value (2.5) is probably justified for fission spectrum neutrons [32]. A source of gamma radiation producing an absorption of 10^{-3} rads/hr in tissue is then equal to 10^{-3} rems/hr. However, a thermal neutron source depositing 10^{-3} rads/hr in the tissue would be equal to 2.5×10^{-3} rems/hr.

The biological effects of nuclear radiation are classified in two categories: (1) somatic effects and (2) hereditary effects. Somatic effects are defined as those related to changes in cells which maintain body functions. Hereditary effects are those causing changes to the portion of the cells responsible for genetic characteristics. Certain tissues are more sensitive to radiation than others. These are the skin, the gonads, the lens of the eye, blood-forming tissues, spleen and lymph nodes, and the gastrointestinal tract.

There are basically two mechanisms by which radiation can interact with cells. These are (1) ionization directly in or in close vicinity to the cell being injured and (2) production of chemical products which are harmful to body functions. Both of these can occur simultaneously, thereby making it difficult clinically to diagnose the cause. Equivalent doses of different types of radiation will produce different biological effects because of the different values of linear energy transfer along the path of the ionizing particle. Table 2-VIII illustrates the term linear energy transfer [33].

Most biological effects are independent of dose rate; for instance, most somatic effects are proportional to the dose, and even a small dose will have some effect. Other biological effects such as the production of chromosome breakages are very dependent upon the dose rate, the breakages produced being greater for high dose rates.

Comparative Risks from Radiation Sources

Man is continually exposed to radiation from natural sources, and collects in the course of an average lifetime about 10 rems of nuclear radiation over the whole body. This dose is contributed by sources both external and internal to the body. One internal source is the radioisotope carbon-14, which deposits approximately 8 mrems/yr in various parts of the body. Potassium-40 is a greater internal contributor, depositing 58 mrems/yr in various parts of the body. The atmosphere contains slight amounts of the radioactive gas radon, which emanates from brick and concrete construction materials.

TABLE 2-VIII

Linear Energy Transfer

Ion pairs/micron (H_2O)	RBE
<100	1
100–200	1–2
200–650	2–5
650–1500	5–10
1500–5000	10–20

This radon gives rise to from 100 to 1000 mrems/yr in the lungs and about 2 mrems/yr in the soft tissues. Externally, the body is exposed to cosmic radiation, which contributes approximately 120 mrems/yr for a sea level location, and 280 mrems/yr for a 5000 ft altitude location. Terrestrial radiations are responsible for 200 mrems/yr. In addition, exposures associated with dental and chest X-rays can contribute as high as 1–2 rems and 0.2–5 rems, respectively, during a single exposure. Fallout from atmospheric nuclear testing contributes approximately 2 mrems/yr.

For handling of radioactive materials the maximum permissible dose has been set to 100 mrems per week or 1.3 rems per 13-week quarter. However, the emphasis is on receiving as small a dose as possible, and to achieve this personnel working with radioactive materials are required by regulations to carefully monitor and record the actual doses received. In this way an awareness of the dangers associated with ionizing radiation is stimulated, and it is quite often found that such personnel because of this knowledge are able to reduce their lifetime radiation dose.

Three types of radiation were listed above as predominant in work with radioactive materials: alpha particles, beta particles, and gamma and X-radiation. The characteristics of each of these from a risk standpoint will now be compared.

Alpha particles are rarely emitted from irradiated electronic equipment, since this equipment is generally composed of low- to medium-density materials. If alpha particles are encountered they will generally be in an energy range which is easily absorbed by a few millimeters of paper, cardboard, or Plexiglas. When emitted externally to the body, these alpha particles are not able to penetrate to the basal layer of the skin, and will at worst produce only skin burns. Care should be exercised, however, to prevent their entrance into the eyes or entrance into the body either through cuts in the skin, or via contaminated food.

Beta particles are almost always encountered when handling irradiated electronic equipment. They are quite penetrating, being able to travel from 3 to 30 ft in air for 0.5 and 3 MeV beta particles, respectively. Their range in denser materials is very much shorter. As an example, a shield made from $\frac{1}{4}$ in. Plexiglas will absorb all beta particles with energies less than 1 MeV, and 1 in. Plexiglas all energies less than 4 MeV. On the other hand, the absorption of the higher-energy beta particles ($> \sim 1$ MeV) gives rise to bremsstrahlung (X-radiation), which is yet more penetrating. The production of bremsstrahlung is a function of the atomic number of the shielding material, such that it is desirable to use a thicker low-density shield rather than a thin high-density shield for the same protection. Beta particles are capable of producing quite severe skin burns, and as with alpha particles extreme care should be exercised to prevent their entrance into the body and into the eyes.

Gamma radiation is always encountered when handling irradiated electronic equipment. Because of the great penetrating power of the gamma rays, it is difficult to shield against them and still be able to handle the irradiated equipment. The best protective measure against this radiation is to limit the exposure time. Protection standards provide that equipment to be utilized when handling the radioactive materials be monitored, so that a check can be kept of the dose received.

Neutrons are rarely encountered by the electronic equipment experimenter, since the test reactors used are by design provided with adequate shielding against these.

Somatic Effects

Somatic effects have been defined earlier as those related to injuries to cells which are concerned with the maintenance of the body functions. Generally distinction is made between acute and delayed effects [32]. The acute effects are defined as those resulting from acute exposure, that is, an exposure where the total radiation dose is received in a short time. The definition for "short" is rather loose, but is generally considered to be a period of less than 24 hr. The delayed effects are defined as those resulting from either an acute or a chronic exposure. Chronic exposure is the receipt of the total radiation dose over a long period of time, such as from natural background radiation. The distinction between acute and chronic exposures is required to allow for the fact that, given time, the body may repair radiation damage if the rate of the cell destruction is not excessive.

For acute exposures 25 to 75 rems will not produce any observable changes in body function behavior. A dose of 75 to 200 rems will produce slight illness, such as vomiting and nausea. For doses from 200 to 600 rems, enough damage to body functions can be incurred that the exposure could have a fatal outcome. For doses of over 600 rems, and especially for doses greater than 1000 rems, a fatal outcome is highly probable [32].

Somatic effects can be summarized as follows:

Immediate effects after exposure to a radiation dose:
1. Skin damage, and necrosis of deep-seated tissues.
2. Impaired function of the nervous system.
3. Changes in the number of white blood cells, a very sensitive barometer of radiation exposure. For relatively low doses (25–200 rems), the number of white blood cells will show immediate response by an abnormal increase during the first days after exposure, followed by a decrease to below normal values.
4. Decreases in the number of platelets, also quite apparent for low doses, and providing an even more accurate barometer for radiation exposure than the white blood cells.

5. Permanent or short-term sterility possibly following irradiation of the gonads.
6. Nausea and vomiting possibly following larger doses (200–600 rems).
7. Disturbances in the gastrointestinal tract function.

Delayed effects, months or years after the exposure:
1. Chronic skin damage.
2. Cataract of the lens of the eye.
3. Leukemia, that is, overproduction of white blood cells.
4. Bone cancer.
5. Premature aging.

One further effect, an indirect one, is that if an individual has disease-carrying bacteria in his blood and his white blood cell number is reduced by radiation exposure, he may succumb to a disease not directly attributable to the radiation exposure.

Different portions of the body show different sensitivity, and there are of course variations in the degree of sensitivity among individuals. It is therefore difficult to generalize as to the specific effect in a particular individual. However, the third and fourth immediate effects are used to evaluate the seriousness of exposure in an individual.

Hereditary Effects

The hereditary effects consist of radiation-induced changes in those components in the reproductive cells which are responsible for mechanisms of heredity. They do not show up in the individual being exposed, but in future generations related to the individual. The effect is a function of dose rate, that is, a high dose rate may be three to five times as effective in producing the effect as a equivalent dose delivered at a low dose rate. The effect is signified by mutations of the genes. Gene mutations produced by radiation do not differ from those occurring naturally. The serious problem presented by the mutations is that the resulting genes are recessive. This means that unless both parents transmit them to the child, the particular characteristic will not be transmitted. This means that a desirable characteristic may be lost for future generations due to the changing of a dominant gene into a recessive gene by mutation.

Radiation Protection Standards

Radiation protection standards should always be observed remembering that there is no such thing as a perfectly safe radiation dose, however small. The Atomic Energy Commission periodically updates protection standards and requirements for safe handling of radioactive materials to reflect changes in national and international criteria regarding such handling. It is the purpose of the following to summarize current requirements and provide information which will enable the reader to comply with these requirements.

The concept of "maximum permissible doses" was established in the period from 1920 to 1930 [34], when it became apparent that occupational hazards were very serious for personnel handling radioactive materials. Ever since that time the tolerance level has been decreased, reflecting an increased understanding of the deleterious effects which ionizing radiation has on the human body. The formulation of the protection standards takes place both on a national and international level. The protection standards have been classified in terms of population classes:

1. Occupational exposure
2. Exposure of special groups
3. Exposure of the population at large

For the purpose of this book only the first class, occupational exposure, will be discussed. This may be divided into protection standards for external and internal body exposure.

External body exposure:

1. The maximum permissible dose for the whole body is 100 mrem per week. It is also specified that a quarterly (13 week) limit of 1.3 rem may be used.
2. The whole body and gonads total lifetime exposure shall not exceed $5(N-18)$ rems, where N is the individual's age in years.

Internal body exposure:

1. For those radioisotopes which are distributed uniformly throughout the body or are concentrated in the gonads, the concentrations are based on an average permissible dose of 100 mrem per week.

2. For those radioisotopes which concentrate in the skeleton, the concentrations are based on a permissible weekly dose to the bone of 0.56 rem.

3. For those radioisotopes which concentrate in single organs of the body other than the gonads, bone, and thyroid, the concentrations are based on a permissible annual dose of 15 rem, and a quarterly dose of 4 rem.

4. For those radioisotopes which concentrate in the thyroid, the concentrations are based on a permissible annual dose of 30 rem and a quarterly dose of 8 rem.

To assure that the above-listed limits are not exceeded and in general to assure that received doses are known, it is important to maintain a careful log of all exposures and utilize radiation monitoring equipment to measure the exposure doses. Two types of monitoring are normally performed: individual monitoring and area monitoring.

Individual Monitoring. The individuals working with radioactive materials should be equipped with two types of monitors: a long-term exposure monitor and a direct read-out monitor. The long-term exposure monitor takes the form of a film badge. Generally carried attached to the front of the individual's clothing, it can also take the form of a finger-ring badge when it is expected that the hands will be more exposed than the rest of the body. The film badge contains a film strip which is partially covered by filters of cadmium, lead, or other material so as to make its response radiation-energy dependent. The use of filters also permits the assessment of doses of different types of radiation. The film when exposed to radiation will darken, and by measurement of the degree of darkening in both the filtered and the unfiltered areas of the film, the exposure dose can be calculated.

The film badge is worn for an extended period of time and would therefore give a somewhat belated warning as to an actual overexposure. To indicate overexposure a direct read-out device is used. It takes the form of a pocket electroscope ionization chamber, which is available in ranges from 100 mrem to 5 rem. It is a small pencil-shaped device and is carried in a shirt pocket. Before use it is calibrated and provides from then on a continuous check on the total exposure dose received up to the limit of its range. These are generally reliable instruments, but mechanical shocks may disturb the readings, and they should never be used alone, rather, only in conjunction with film badges.

Area Monitoring. Laboratories in which radioactive materials are handled should be equipped with an area monitor. There are many types, but two are most generally used, the ionization chamber and the Geiger-Müller counter.

The ionization chamber normally comes with range switching because of the large range of dose rates covered. It measures directly the electric charge released in a certain volume of air by the radiation and thus provides the most direct method of measuring dose, to an accuracy of about ±3%.

The Geiger–Müller counter, as the name implies, counts particles. Specifically, in the case of gamma radiation it counts the secondary electrons produced as the gamma radiation interacts with the wall materials of the Geiger–Müller tube. Beta particles can be counted directly if a thin window is provided for their entrance in the G–M tube. The G–M counter is a sensitive device and excellent for measuring low dose rates. It has a disadvantage in that it can saturate. When this happens, the meter reading returns to zero, giving a erroneous indication of actual conditions. It is therefore advisable not to use the G–M counter alone, but only in conjunction with an ionization chamber.

REFERENCES

Radiation Types

1. V.A.J. VanLint and E.G. Wikner, IEEE Transactions of Nuclear Science, NS-10 (January 1963).

Semiconductor Physics

2. D. Dewitt and A. L. Rossoff, Transistor Electronics, McGraw-Hill Book Company (New York), 1957.
3. W.R. Beam, Electronics of Solids, McGraw-Hill Book Company (New York), 1965.

Permanent Displacement Effects

4. G. Bemski, J. Appl. Phys. 30:1195 (1959).
5. G.D. Watkins and J.W. Corbett, J. Appl. Phys. 30:198 (1959) and Phys. Rev. 121(4):1001 (February 1961).
6. D. Billington and J.H. Crawford, Radiation Damage in Solids, Princeton University Press (Princeton, New Jersey), 1961.
7. J. Loferski, "Analysis of the Effect of Nuclear Radiation on Transistors," J. Appl. Phys. 29:35 (1958).

8. J. W. Easley and J. A. Dooley, "Irradiation of Germanium Alloy Transistors," AGET Symposium (February 1957).
9. G. C. Messenger and J. P. Spratt, "The Effect of Neutron Irradiation on Germanium and Silicon," Proc. IRE 46:1038 (1958).
10. J. R. Bilinski, "Selecting Transistor for Radiation Environment," Electronics (December 1959).
11. J. R. Bilinski et al., "Proton Neutron Damage Equivalence in Semiconductors," IEEE, NS-10, p. 71 (November 1963).
12. L. B. Gardner and A. B. Kaufman, "Semiconductors in a Hyper-Nuclear Environment," WESCON show paper 20/2 (August 1961).
13. D. Hendershott et al., "Transistor Selection Technique," General Electric Report ETR-8234-008.
14. Lindmayer and Wrigley, "Beta Cutoff Frequencies of Junction Transistors," Proc. IRE (February 1962).
15. G. C. Messenger et al., "Displacement Damage in MOS Transistors," IEEE Nuclear Science Group Meeting (July 1965).
16. S. K. Manlief, "Neutron Induced Damage to Silicon Controlled Rectifiers," IEEE, NS-11 (November 1964).

Ionization Effects

17. H. L. Olesen, "Telemetry Circuits in a Pulsed Radiation Environment," Proc. National Telemetry Conference (1963) and IEEE, NS-11 (April 1964).
18. V. S. Vavilov, "Effects of Radiation on Semiconductors," Consultants Bureau (New York), 1965.
19. W. VanRoosbroeck, J. Appl. Phys. 26:380 (1955).
20. J. L. Wirth and S. C. Rogers, "The Transient Response of Transistors and Diodes to Ionizing Radiation," IEEE, NS-11 (November 1964).
21. E. A. Carr, "Transient Radiation Effects on Transistors," IEEE, NS-11 (November 1964).
22. J. J. Ebers and J. L. Moll, "Large Signal Behavior of Junction Transistors," and J. L. Moll, "Large Signal Transient Response of Junction Transistor," Proc. IRE 42 (December 1954).
23. C. Rosenberg et al., "Charge Control Equivalent Circuit for Predicting Transient Response," IEEE, NS-10 (November 1963).
24. J. Raymond et al., "Radiation Effects in MOS Transistors," IEEE, NS-12 (February 1965).
25. D. S. Peck et al., "Surface Effects of Radiation on Transistors," Bell System Technical Journal (January 1963).
26. H. L. Steele, Jr., "Effects of Gamma Radiation on Transistor Parameters," Proceedings of the Transistor Reliability Symposium (September 1956).
27. R. R. Blair, "Surface Effects of Radiation on Transistors," IEEE, NS-10 (November 1963).
28. M. M. Weiss and W. P. Know, "The Effects of Gamma Radiation on Glass Coated Silicon Transistors," IEEE, NS-10 (November 1963).
29. J. C. Peden et al., "Radiation Surface Effects in Silicon Planar Transistors," paper presented at IEEE Radiation Effects Conference (1964).
30. C. W. Bostian and E. G. Manning, "The Selection of Transistors for Use in Ionizing Radiation Fields," IEEE, NS-12 (February 1965).

Effects on Man

31. NCRP Report No. 11, NBS Handbook 52 (1953).

32. S. Glasstone, "The Effects of Nuclear Weapons," U.S. Government Printing Office, Paragraph 11.88, Revised Edition (February 1964).
33. E. L. Saenger, "Medical Aspects of Radiation Accidents," U.S. AEC, U.S. Printing Office (1963).
34. L. S. Taylor, "History of the International Commission on Radiation Protection," Health Phys. 1 (1958).

BIBLIOGRAPHY
Radiation Types

Billington, Douglas, and James H. Crawford, Radiation Damage in Solids, Princeton University Press (Princeton, N.J.), 1961.

Cohen, B. L., "Nuclear Orbital Structure," Science and Technology (November 1963).

De-Shalit, Amos, and Igal Talmi, Nuclear Shell Theory, Academic Press (New York), 1963.

Dienes, G. J., and G. H. Vineyard, Radiation Effects in Solids, Interscience Publishers (New York), 1957.

VanLint, V. A. J., and E. G. Wikner, "Correlation of Radiation Types with Radiation Effects," IEEE Transactions of Nuclear Science, NS-10, No. 1, p. 80 (January 1963).

Semiconductor Physics

Biondi, F. J. (ed.), Transistor Technology, D. Van Nostrand Company, Inc. (Princeton, N.J.), 1958, Vol. II.

Bitter, F., Current, Fields, and Particles, John Wiley & Sons, Inc., (New York), 1956.

Effects

Billington, Douglas, and James H. Crawford, Radiation Damage in Solids, Princeton University Press (Princeton, N.J.), 1961.

Blatz, Hanson, Introduction to Radiological Health, McGraw-Hill Book Company, Inc. (New York), 1964.

Browning, E., Harmful Effects of Ionizing Radiations, Elsevier Publishing Company (Amsterdam), 1959.

Glasstone, S., "The Effects of Nuclear Weapons," U.S. Government Printing Office.

Hanson, G., "Semiconductor Device Irradiation," Texas Instrument Report (July 1964).

Harwood, J. J., et al., Effects of Radiation on Materials, Reinhold Publishing Corp. (New York), 1958.

Saenger, E. L., "Medical Aspects of Radiation Accidents," U. S. Atomic Energy Commission, U. S. Government Printing Office (August 1963).

Schubert, J., and R. E. Lapp, Radiation: What It Is and How It Affects You, Viking Press (New York), 1957.

Strayhorn, T. R., "Radiation Heating," Space/Aeronautics 39 (3):111 (March 1963).

Sulit, R. A., et al., Principles of Radiation and Contamination Control, Vols. I and II, U.S. Naval Radiological Defense Laboratory (1959).

Vavilov, V. S., "Effects of Radiation on Semiconductors," Consultants Bureau (New York), 1965.

Watson, B. B., The Delayed Effects of Whole-Body Radiation, John Hopkins Press (Baltimore, Md.), 1960.

The combined works of the Radiation Effects Information Center at Battelle Memorial Institute, Columbus, Ohio.

CHAPTER 3

Radiation Shielding

INTRODUCTION

Shielding is needed for protection of components and devices against many forms of radiation. There are two primary types of shielding, passive and active, each so named according to its manner of damping the motion of the fundamental particles. The theory of these shields is given in this chapter, and some of the available data are presented.

Passive shielding employs physical materials to absorb, or decelerate and absorb, radiation. In the past, the weight of materials needed to furnish appreciable protection from severe radiation prevented the use of passive shielding in flight equipment. With the advent of microelectronics, where circuit volume is being shrunk to a tenth or a hundredth of its present size, passive shielding is becoming feasible for more applications. Design data are presented in this chapter for several shield materials.

Because of the large weight necessary for effective passive shielding of whole vehicles, active shielding methods are also being considered. Two types of active shielding are possible: electrostatic and magnetic. The features of each will be discussed in more detail below.

PASSIVE SHIELDING

Neutron Bombardment

The effects of neutrons bombarding electronic components can be very serious, and the requirements of shielding must be considered.   Normally, encapsulating materials are not dense enough to stop a significant number of damaging neutrons. The elimination of gases or air, which reduces ionization effects, does not reduce the effect of neutron bombardment. The best technique but most impractical method for slowing the fast neutrons is shielding of the system.   Concrete is normally used in a ground reactor.

In a space vehicle, shielding against fast neutrons with energies greater than 10 keV is very difficult.   Since the neutrons have no electrical potential, they pass unperturbed through any material and, except when in a direct collision with a nucleus, do not lose their energy.   The attenuation of the neutrons involves several different phenomena.

Typically a four-layer shield is used to stop fast neutrons. The "fast" neutrons are slowed to the moderately fast range by a layer of inelastic scattering material, such as steel. These moderately fast neutrons are slowed down further, to the "slow" or "thermal" range by a layer of hydrogen-containing material such as LiH.   The thermal neutrons can now be absorbed in a layer of boron, which has a large cross section for such neutrons. However, the excited boron nuclei emit gamma rays, and a backup layer of high-z (high atomic number) material is provided to stop these and to serve as a heat sink.

Obviously, such a combination shield consisting of steel, hydrogenous material, boron, and steel becomes very cumbersome and not at all practical for space or re-entry vehicle applications. At best the hydrogenous material may be included to reduce the total dose of neutrons by a factor of 2 to 3.

Gamma Radiation

Gamma radiation can cause transient effects since the initial radiation effect from a pulsed source can be very intense. The primary effect is not permanent damage.   The gamma pulse will cause serious problems due to ionization of gases and the surrounding air, as well as of organic and semiconductor materials. Some of the possible solutions will be discussed.

Ionization of air and gases can cause serious leakage paths for the flow of current in electronic equipment. In a transistor the effect can be reduced by sealing the transistor in a vacuum, or by filling the transistor can with a material that will not be easily ionized. The effect of the ionization at the terminals of a transistor or any other component can be reduced by encapsulating the critical areas.

Up to now little information has been available on the effect of transient radiation on wires that are in a bundle or cable. A recent experiment had a typical re-entry vehicle subsystem exposed with the associated wire bundles [1]. At the high dose rate of 10^{12} rads (C)/sec, no additional effects were noticed which could be blamed on cabling.

The air ionization created by the high-intensity gamma radiation is handled in two ways. In transistorized circuits where heat generation is low, it is possible to encapsulate the circuits with conformal materials, such as silicone rubber,

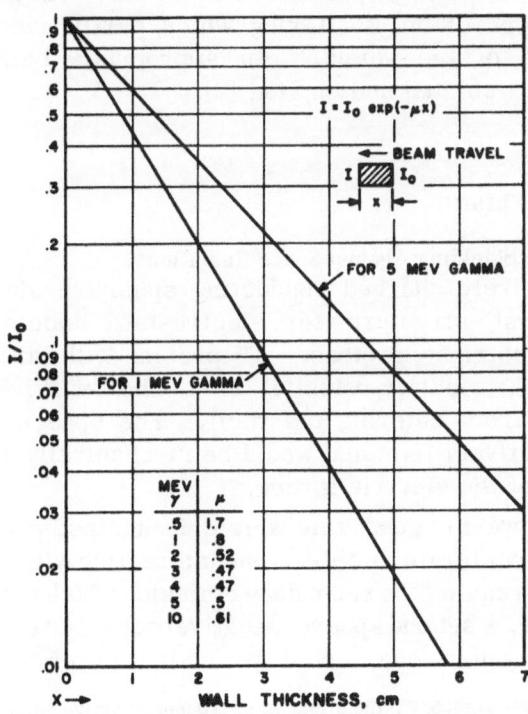

Fig. 3-1. Gamma dose rate attenuation by lead (build-up factor is not included).

thereby eliminating air pockets which would support air ionization across terminals of different potentials. In vacuum tube circuits, TIMM circuits, and other circuits where heat generation is great, it has been found necessary to evacuate the air, and as experimental data indicate, a vacuum of $1\,\mu$ Hg or less is required to achieve full protection against air ionization.

When considering the weight limitations put on a re-entry vehicle, direct passive shielding against the gamma radiation is difficult. This is so because only materials of high atomic numbers will effectively attenuate the gamma rays. Three types of interaction will cause this attenuation: (1) Compton effect, (2) photoelectric effect, and (3) pair production. Figure 3-1 indicates the amount of shielding required for a typical gamma ray of energy 5 MeV, and it is obvious how heavy an effective shield a typical re-entry vehicle would have to carry. The figure is strictly valid for thin shields; with thick shields, there is appreciable scattering which partly cancels the effectiveness of the shield. An appropriate build-up factor should be calculated and applied [2].

ACTIVE SHIELDING

Electrostatic Shielding Against Space Radiation*

A positively charged conducting spherical shell provides the simplest structure for electrostatic shielding against incoming isotropic protons. All protons with initial energies less than the electric potential on the shell will be completely prevented from entering the shell. The spherical shell can be comparatively light and would be mechanically stable under the action of the electric forces.

But, however good this shield is against protons it will obviously accelerate any electrons interacting with the vechicle, thereby increasing the secondary emission which these produce. In principle, a second sphere, negatively charged and concentric inside the outer one, could be added to provide shielding

*Taken in part from S. W. Kash and R. F. Tooper, "Active Shielding for Manned Spacecraft," Astronautics, Sept. 1962.

against the electrons. Some means would have to be provided to keep the two spheres concentric without reducing the potential difference between them. This concentric arrangement would be dynamically unstable and a slight off-center displacement would result in large forces tending to draw them quickly together.

The potentials on the spheres are also limited by the capacity of available electrostatic generators. Present-day generators cannot provide the electric potentials necessary to stop the more energetic electrons or protons in space. Charge leakage between the spheres would result in a serious power loss.

In certain special situations electrostatic shielding may offer a weight saving over passive shielding, but has other disadvantages as pointed out. It may be possible to combine passive shielding against electrons with electrostatic shielding against protons, and achieve a system lifetime compatible with requirements.

Magnetic Shielding Against Space Radiation*

Magnetic shielding has become practical with the recent discovery of "hard" superconducting materials, such as niobium–tin and niobium–zirconium. These alloys attain superconductivity in magnetic fields of the order of 100,000 G and can carry high current densities. The high current-carrying capacity of these hard superconductors — in sharp contrast to that of "soft" superconductors, such as lead and tin, in which all of the current must be carried in a thin layer on the surface — seems to be derived from the fact that the currents flow within the body of the conductors in superconducting filaments.

To appreciate the advantage a magnetic shield offers over a passive one for charged particles, we may consider the circular deflection of a proton moving in a transverse magnetic field. For particle kinetic energies smaller than the particle rest energy (about 1 BeV for protons) the radius of the circle of deflection (in cm) is proportional to the half-power of the energy:

*See the footnote on p. 122.

$$R = 2.88 \times 10^5 \left(\frac{\sqrt{E}}{B} \right)$$

where E is the proton energy in MeV and B is the magnetic field strength (G).

An important parameter when designing a magnetic shield is known as the Stoermer unit or the Stoermer radius. This is the radius of the circular particle orbit in the equatorial plane of a magnetic dipole field and is given by the following expression:

$$C_{ST} = \sqrt{\frac{ea}{m_0 cv}} \sqrt[4]{1 - \frac{v^2}{c^2}}$$

In terms of the energy of the particle the Stoermer radius is given by the following expression:

$$C_{ST} = \left\{ \left(\frac{ea}{E + m_0 c^2} \right) \frac{1}{\sqrt{1 - [m_0^2 c^4/(E + m_0 c^2)^2]}} \right\}^{1/2}$$

The graph in Fig. 3-2 shows the Stoermer unit as a function of proton energy at various magnetic moments. Figure 3-3 shows the shielding which will be provided in a toroidal fashion around the vehicle by a magnetic dipole shielding system.

SHIELDING AGAINST LONG-TERM SPACE RADIATION*

The matter of shielding against space radiation becomes important for space vehicles which have to spend long-term missions in the space radiation regions and still maintain their operation for many years. During their stay in the proton and electron fields of the Van Allen belts and those fields due to solar flares the vehicles will accumulate radiation dosages that will eventually exceed the tolerance of the electronic circuits aboard the vehicles. Among the type of vehicles that will be expected to operate with high reliability for many years in orbit are communication satellites, weather satellites, astronomical satellites, and space stations.

*Taken in part from S. W. Kash and R. F. Tooper, "Active Shielding for Manned Spacecraft," Astronautics, Sept. 1962.

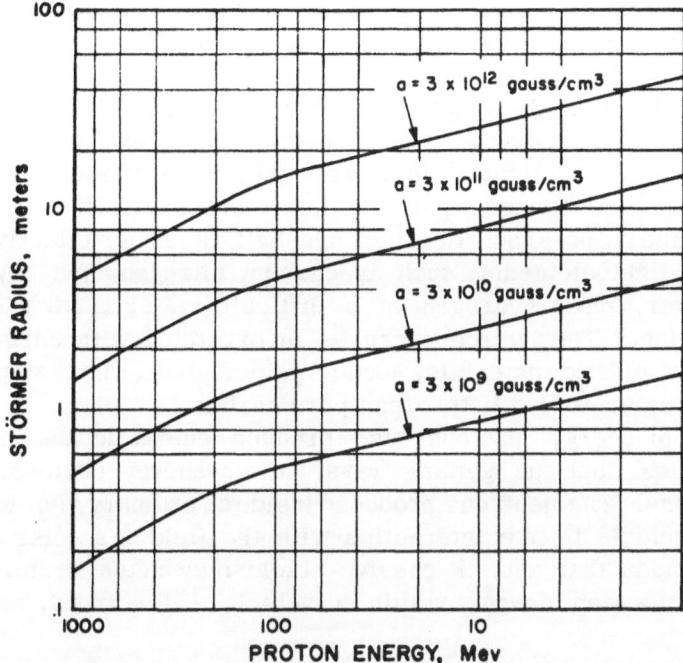

Fig. 3-2. Shielding against moving charged particles by high-intensity magnetic fields (after Astronautics, September 1962).

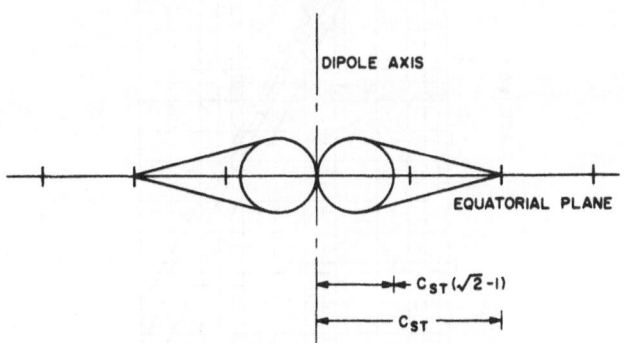

WITHIN $C_{ST}(\sqrt{2}-1)$: COMPLETELY SHIELDED REGION
WITHIN C_{ST} : PARTIALLY SHIELDED REGION

Fig. 3-3. Shield pattern established by magnetic dipole against charged particles (after Astronautics, September 1962).

All three methods of shielding under consideration at the present time are feasible. In passive shielding the energy of the incoming particles is dissipated by multiple collisions, primarily with the electrons within the materials. The graph in Fig. 3-4 shows the range of protons in several elementary materials. The lighter elements provide more effective shielding per unit mass than the heaviest ones. Hydrogen, in particular, is about two and one-half times as effective as other light elements, such as carbon, nitrogen, and oxygen. However, pure hydrogen is a difficult material to use for shielding. The hydrogen can be incorpotated with carbon in various plastic materials, such as polyethylene, whose stopping power is somewhat better than pure carbon.

High-energy protons also produce energetic secondary particles such as gamma rays, lower-energy protons, and neutrons. The neutrons produced inside an ordinary aluminum-skin vehicle by the interaction with the field of a solar flare may mean that a thick passive shield may actually increase the ionization dosage within a vehicle. Electrons, having

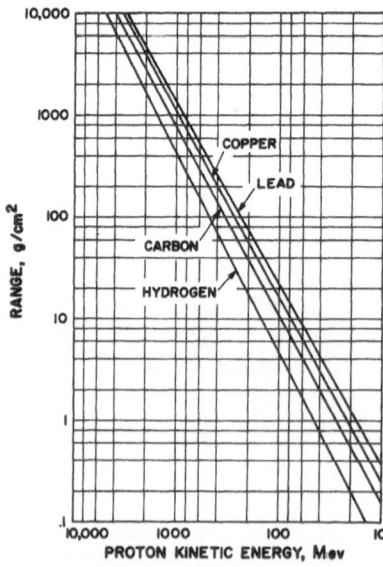

Fig. 3-4. Proton penetration of passive shield material (after Astronautics, September 1962).

energies usually lower than protons and a much smaller range, pose less of a shielding problem. When stopped, however, the electrons produce secondary X-rays more penetrating than the electrons themselves.

Estimates of passive shield requirements have been made for various solar events and assumed dosage rates. Shield weights of the order of a few tons may be necessary for moderate solar flares. For example, 334 g/cm^2 of polyethylene are required to shield against 1 BeV protons. For missions lasting only a few days or weeks, however, considering the probability of solar flare occurrence, the radiation dose can be reduced to an acceptable level by a relatively small amount of shielding. An analysis of this situation for a typical spacecraft is given below.

Radiation Levels Received Within a Typical Spacecraft*

The hazard resulting from exposure of spacecraft to space radiation elements has been evaluated using the kinetic energy spectra models already discussed. The spacecraft configuration now considered is typical of those designed for lunar missions. Its forward section is conical, having a diameter at the base approximately 160 in. and a length of approximately 130 in. The spacecraft is composed of a varying thickness of ablative material on the outer surface, a titanium and aluminum pressure shell, heat insulation, and an interior structure with internal equipment. The average material thickness of the vehicle is 13.86 g/cm^2.

At a dose point located in the vicinity of the astronaut's body, the radiation dose rates from the May and February (see Table 3-I) solar events and the inner radiation belt protons at peak intensity are 1.5, 36, and 5 rem/hr, respectively. The primary proton radiation doses for the May and February solar events and for the inner radiation belt have been evaluated at a dose point located near a crewman's seat. Computed in the manner previously discussed, the primary proton radiation doses are as follows: (1) 46 rem for the May event, assuming the peak intensity of the proton flux is constant for 30 hr and

*Taken in part from A. J. Beck and E. L. Divita, "Evaluation of Space Radiation Doses Received within a Typical Spacecraft," ARS Journal, Nov. 1962.

TABLE 3-I

Comparison of Dose Rate Contributions Due to C^{12} (p, xn) and C^{12} (p, xγ) Reactions

Environment source	Target and process	Dose rate (rad/hr)
May 10, 1959, solar flare	C^{12} (p, xn)	3.7×10^{-2}
	C^{12} (p, xγ)	9.1×10^{-4}
February 23, 1956, solar flare	C^{12} (p, xn)	2.9×10^{-2}
	C^{12} (p, xγ)	1.1×10^{-4}
Van Allen protons at peak intensity	C^{12} (p, xn)	4.2×10^{-3}
	C^{12} (p, xγ)	9.7×10^{-6}

then decays immediately to zero; (2) 36 rem for the February event, assuming the peak intensity immediately decays as t^{-2}; and (3) 5 rem for the inner radiation belt, assuming the peak intensity of the proton flux is constant for 1 hr. The primary proton radiation doses for the same events, using an aluminum sphere with a 13.86 g/cm^2 average thickness, are 8.5 rem for the May solar event, 30 rem for the February solar event, and 4 rem for the inner radiation belt.

A comparison of the radiation doses obtained by the two methods shows that the values obtained using spherical models with average thicknesses of materials give results that are generally too small. The difference strongly depends on the proton energy spectrum. For example, the results obtained from the detailed analysis of the vehicle differ by a factor of 5.41 from the aluminum sphere analysis for the May event, whereas they differ by factors of 1.2 and 1.25 for the February event and the inner radiation belt, respectively.

The secondary neutron energy spectrum within the space-craft for the May solar event tended to flatten out in the energy range from 1 to 10 MeV, due to different thresholded energies of the various reactions. The radiation dose from secondary neutrons produced by protons from the May flare is 9 rem, whereas the neutron dose from the February flare and the inner radiation belt at peak intensity are 5 and 0.54 rem, respectively. These values are for the secondary neutrons only.

Table 3-I shows the relative importance of secondary gamma radiation from $(p, x\gamma)$-type reactions. Here the percentage contribution for the C^{12} $(p, x\gamma)$ reaction is, at times, 3% when compared with the C^{12} (p, xn) reaction dose rate. This comparison does not dismiss the $(p, x\gamma)$ reaction radiation but at least implies that its contribution would be considerably less than the (p, xn) reaction radiation.

The secondary bremsstrahlung contributes quite largely to the total dose rate received within the Van Allen radiation belts. The doses corresponding to the three sources mentioned before are 1.044, 0.765, and 0.2404, respectively. For this typical spacecraft, the Van Allen electrons do not contribute to the dose within the spacecraft.

Conclusions

The total radiation dose received during a space flight depends primarily on the radiation resulting from solar flares. The occurrence of two solar flares — one similar to the May, 1959, event and one similar to the flare of February, 1956 — would result in a total radiation dose of about 98 rem. As discussed previously, solar flares similar to the May event occur three to four times per year during the period of maximum solar activity, whereas a solar flare similar to the February event occurs only once every two years. Thus, the probability of receiving a radiation dose not exceeding 98 rem from the two solar events during short duration missions is large (i.e., 0.997 for a 7-day mission and 0.990 for a 14-day mission).

The dose for transit through the Van Allen belts has been evaluated, assuming a dose equivalent to 1 hr exposure at peak intensity. However, detailed evaluation based on typical lunar trajectories indicates that the total proton dose, including secondaries, is about 0.6 rem, whereas the electron-bremsstrahlung dose is about 0.2 rem. Assuming, finally, a cosmic ray dose of 2.8 rem, the total 14-day mission dose comes to about 102 rem. In addition, 5.5 g/cm^2 of aluminum shielding could reduce the radiation dose to 59 rem, 31 rem of which result from the February solar flare, an event with very small probability of occurrence. Therefore, considering the occurrence probability of solar flares, radiation doses received during a typical mission can be reduced to an acceptable level with radiation shielding.

REFERENCES

1. H. H. L. Olesen, "Telemetry Circuits in a Pulsed Radiation Environment," IEEE, NS-11 (April 1964).
2. B. T. Price, et al., Radiation Shielding, Pergamon Press, Inc. (New York), 1957.

BIBLIOGRAPHY

Beck, A. J., E. L. Divita, and S. L. Russak, "Evolution of Space Radiation Safety Procedures in the Design and Operation of Some Early Manned Lunar Vehicles," in: C. T. Morrow et al. (eds.), Proceedings of 6th Symposium on Ballistic Missile and Aerospace Technology, Academic Press Inc. (New York).

Beck, A. J., and E. L. Divita, "Evaluation of Space Radiation Doses Received Within a Typical Spacecraft," ARS Journal (November 1962).

Beischer, D. E., "Human Tolerance to Magnetic Fields," Astronautics (March 1962).

Kash, S. W., and R. F. Tooper, "Active Shielding for Manned Spacecraft," Astronautics (September 1962).

Price, B. T., et al., Radiation Shielding, Pergamon Press, Inc. (New York), 1957.

Störmer, C., The Polar Aurora, Clarendon Press (Oxford), 1955.

Tooper, R. F., and W. O. Davies, "Electromagnetic Shielding of Space Vehicles," IAS Paper No. 62-156 (June 1962).

Tooper, R. F., and W. E. Zagotta, "Design of a Superconducting Solenoid for Magnetic Shielding," Armour Research Foundation Report ARF 1196-9.

CHAPTER 4

Experimental Facilities

RADIATION EXPERIMENTATION

Many radiation facilities exist in the United States and most of these are available for radiation testing. Facilities outside this hemisphere amount only to about 50% of the radiation facilities in the world, and none of those are unique in the sense that they cannot be found in this hemisphere. As a consequence the following compilation will concentrate only on the facilities in the United States and only describe those which, it is felt, will be of general help to an electronic equipment experimenter.

To establish the type of facility that is needed for experimentation, the environment in which the equipment must operate must be evaluated. There are basically three types of environment with which the designer will be concerned: (1) radiation in space, (2) radiation from nuclear reactors or nuclear propulsion systems, and (3) radiation from a weapon burst.

The radiation facility for simulation should be chosen according to these expected environments. A list of corresponding environments and radiation sources follows:

1. Radiation in space: steady-state gamma and high-energy proton sources.
2. Radiation from nuclear reactors and nuclear propulsion systems: steady-state gamma, steady-state gamma–neutron, and pulsed gamma–neutron sources.
3. Radiation from weapon bursts: short-pulse gamma (Linac and X-ray), short-pulse electron (Linac), and pulsed gamma–neutron sources, and actual nuclear weapon tests and RFI facilities (electromagnetic).

TABLE 4-I

Steady-State Gamma Facilities *

Output [ergs/g(C)-hr]	Source	Sample environment	Sample size	Other considerations	Location
2×10^7 ($E=1.7$ and 1.33 MeV)	Co^{60}	Water	4 in. φ × 7 in.	1. Temperature environment can be varied 2. Easy contractual arrangements 3. Cost: short term, $70; long term, <$5/hr 4. Availability: 2 weeks	Battelle Memorial Institute 505 King Ave. Columbus, Ohio
3×10^7	Co^{60}	Air or water	3 in. φ × 40 in.	1. Pressure and atmosphere environment variable	Dr. C. G. Collins General Electric Co. Cincinnati 15, Ohio
4×10^7	Co^{60}	Air	13 in. φ × 12 in.	1. Television available 2. For GE use	Mr. Black General Engineering Lab General Electric Co. Schenectady, N.Y.

3×10^8 ($E = 0.1-2.9$ MeV)	Spent nuclear fuel cells	De-ionized water	3 in. $\phi \times 36$ in.	1. For GE use	Vallecitos Atomic Lab General Electric P. O. Box 846 Pleasanton, Calif.
10^9	Co^{60}	De-ionized water	$^7/_8$ in. $\phi \times 6$ in.	1. For GE use	Vallecitos Atomic Lab General Electric P. O. Box 846 Pleasanton, Calif.
2×10^7	Co^{60}	Air	$2^1/_8$ in. $\phi \times 5^1/_2$ in.	1. Designed for electronic experiments 2. Cost: \$100/day	Dr. H. L. Wiser Radiation Systems Dept. Hughes Aircraft Co. Culver City, Calif.

*Supplementary information can be found in D. J. Hamman and W. H. Veazie, Jr., Radiation Effects Information Report No. 31, Part I, Battelle Memorial Institute (Columbus, Ohio), September 1963.

STANDARD DATA SHEET

Material or Part:
Manufacturer's Identification:
Military Specification:
Composition:
Past History:
Size of Sample:

Property	Initial Value	Per Cent Change, + or -	Gamma, ergs g^{-1}(C)	Description of the Radiation Field				Total Absorbed Energy, ergs g^{-1}	Exposure* Time, hours	Exposure Temp, °F	Other Environmental Conditions	Reference
				Thermal-Neutron Integrated Flux, (nv$_0$)t	Fast-Neutron Integrated Flux, n cm^{-2}	Neutron Energy, Mev						

*Note: If the exposure rate is not constant during the exposure time, more detailed information should be included.

Fig. 4-1. Standard data sheet. [Recommended by the Radiation Effects Information Center of Battelle Memorial Institute, D.C. Jones et al., Radiation Effects Information Center Report No. 35, Battelle Memorial Institute (Columbus, Ohio), August 1964.]

TABLE 4-II

Steady-State Gamma–Neutron Facilities

Output	Source	Sample environment	Other considerations	Location
1×10^{13} neutrons/cm^2-sec (thermal) 10^{13} neutrons/cm^2-sec (>10 keV) 4.3×10^9 ergs/g (C)-hr	2 MW pool reactor	Water-tight containers, sample containers, 8-in. flux tubes	1. Operates continuously for 12 days at a time 2. Easy contractual arrangements 3. Cost:short term, $50; long term, <$125/hr 4. Availability:2 weeks	Mr. J. W. Chastain, Jr., Battelle Memorial Institute, 505 King Avenue, Columbus, Ohio (Contact A. Plummer)
2×10^9 neutrons/cm^2-sec (>10 keV) gamma dose not known	30 kW homogeneous reactor	Air	● Operates 8 hr a day ● Sample room 4 ft cube ● For GE use	Atomic Power Equipment Dept., General Electric Co., 2151 S. First Street, San Jose, Calif.
10^{15} neutrons/cm^2-sec (>10 keV) 10^{11} ergs/g (C)-hr	30 MW tank type reactor	Air	● For GE use	Atomic Power Equipment Dept., General Electric Co., San Jose, Calif.,
3.6×10^{12} neutrons/cm^2-sec (>10 keV) 5.4×10^9 ergs/g(C)-hr	250 kW pool type TRIGA MK I reactor	Water-tight container in F-ring	1. Easy contractual arrangements 2. Availability:4 weeks 3. Sample size <1.4 in. dia.	Dr. A. T. McMain, John Jay Hopkins Lab., General Atomic, P. O. Box 608, San Diego, Calif.

Tables 4-I and 4-II present summaries of the steady-state gamma and steady-state gamma—neutron facilities. Figure 4-1 gives a standard data sheet.

EXPERIMENTAL CONSIDERATIONS

The word "test" means a trial which can result in one of two alternative decisions. A test results in the acceptance of one hypothesis and the rejection of an alternate hypothesis. The experimentation provides the evidence upon which the rejection or acceptance of a test is based. But experimentation is a costly procedure, and good justification should be brought forward that a certain experiment is at all necessary. It may be that, by proper evaluation of data from previous experimentation on similar events, a trend emerges which by application to the problem at hand eliminates the need for experimentation or at least reduces the size of the experiment. Serious consideration should also be given to the arrangement of the experiment, such that every possible piece of information can later be deduced from the data gathered. An experiment should always be instrumented for measurement of various events that may not be of immediate interest but may provide data to minimize future experimentation.

If by analysis of the radiation problem a decision is made to perform an experiment to clarify doubtful issues, the next step is to choose a radiation facility which will most effectively yield data of value. The choice of radiation facility will greatly depend upon the type of environment it is desired to simulate. The list of facilities given later is arranged in order to make this choice as quickly as possible. Emphasis has been placed on the use of readily available facilities in order to minimize the often sticky red tape the experimenter has to cut in order to get at his ultimate goal — the testing. It is also important to decide whether the experiment should be tested actively, i.e., with power applied, or passively. The conditions of the final systems in the space vehicles will in most cases determine which is preferred.

It should be emphasized that in some cases the equipment in the experiment may not be retrievable due to induced radioactivity of long half-lives. In most pulsed reactor tests only a few days are required for a typical electronic circuit to cool down to a safe level of handling, whereas experiments performed over a long period in steady-state reactors require several months to cool down. It should be mentioned at this point that an AEC license for handling radioactive materials is required, and those workers involved with the handling of these materials are kept on a special access list which is kept up-to-date by cognizant personnel.

Experimental Design

The design of the experiment is entirely dependent upon the following factors: (1) the hypothesis to be proved or disproved, (2) the risks associated with a wrong decision, and (3) the parameters that describe the device referred to in the hypothesis. It is imperative to know what parameters describe the system to be evaluated. We must know which parameter will show the changes we wish to detect. Further, we should estimate the magnitude of change expected so that we can set up the measuring equipment. Knowing these things we must work to control the experiment so that unknown variables do not disturb the results.

In controlling radiation experiments several things must be kept in mind. What types of conditions will the equipment experience, in addition to the nuclear radiation? These will be heat, humidity, temperature change, and physical stresses. In order to properly evaluate the radiation effect, all these other effects must be measured so that a differentiation can be made during data reduction. A compensation could also be included in the experiment, by the addition of a control experiment. This consists of a parallel experiment run in a non-nuclear environment, but closely related to the nuclear experiment as far as other environments are concerned.

Even using the best available measuring technique and cable for the experiment, some radiation effect on the cable is unavoidable; therefore, a set of control cables must be included in the experiment in order to monitor these effects.

Test Monitoring

From the description and location of the available facilities, the varied exposure requirements, and the long delay expected in obtaining the planned reactors, it becomes evident that the piece parts electronic monitoring equipment associated with this program is best made completely mobile. Also, the equipment should be shock-mounted to withstand transportation environments during trips over thousands of miles. The equipment must also be completely reliable, since limitations in on-site pre-experimental time virtually preclude the possibility of troubleshooting for equipment malfunctions at the site. An example of the type of fully equipped vehicle which is being used in radiation effects experiments by the U.S. Army Signal Corps is the Radiation Effects Mobile Laboratory, one of the most complex instrumentation vehicles now in use. With the instrumentation in the REML, it is possible to monitor 90 channels of dynamic information at one time. This instrumentation system is needed because of the large number of devices which are exposed during each experiment. The number of devices is based on the minimum sampling of devices which will provide statistically valid information.

To determine the effects of nuclear radiation on a single component, such as an electron tube or transistor, all associated circuitry must be completely shielded from the radiation field. The intense radiation fields from a pulse reactor necessitate extreme precautions to preclude effects on monitoring equipment and associated circuitry. Transient effects noted during some early field experiments at pulsed nuclear reactors were later found to be attributable to radiation effects on an air capacitor oscillator, which was part of the instrumentation equipment [2]. Extreme care must be taken to eliminate all extraneous phenomena, such as cable effects and air ionization, which cause, among other things, shunt leakage. The latter effect can be reduced by potting the test chassis with paraffin or other solid dielectrics.

Dosimetry is also an area in which deficiencies exist. The radiation levels to which a device is exposed must be known before any correlation or analysis is possible. Many times it is impossible to determine the dose or dose rate at the exposure sample and, therefore, only mean values are available, necessitating extrapolation which adds more errors. Dosimetry is rapidly advancing, and devices such as the Magnesium Oxide Radiation Detector (MgO-RAD), developed by R. G. Saelens, as well as the Semi-Rad, developed by Dr. S. Kronenberg and H. Murphy of USASRDL, are helping to alleviate some of the problems associated with gamma dose rate measurements.

Difficulties have also arisen in correlating transient changes in electron device operation to the type of incident radiation, i.e., neutron and gamma, because of lack of adequate rate dosimetry. The size of many detectors precludes measurement of radiation levels at the same position as the exposure sample and, therefore, one must extrapolate to find the exact dose level to which the part has been exposed.

Spectrum Determination

The different spectra of a number of nuclear sources are listed in Table 4-III. An examination of the table will quickly reveal the importance of accurately reporting the neutron spectrum, so that later correlation can be made with radiation test data from a different neutron source. The table gives the fraction of the 10 keV to 18 MeV dose and the 1 MeV to 18 MeV dose when a total dose is measured between 4×10^{-7} and 18 MeV.

The >10 keV to >3 MeV ratio for the Ground Test Reactor is approximately 4 and for the Godiva it is approximately 7. For a $1/\dot{E}$ plus fission spectrum the ratio is approximately 4.5. The 10 keV to 3 MeV ratio is more commonly used, since two dosimeter points (Pu and S) are generally readily available for 10 keV and 3 MeV.

Referring to Chapter 2, the reader will find that the radiation damage is a function of neutron energy. It is therefore extremely important to ascertain with as good accuracy as possible the neutron spectrum.

Characteristics of Ground Test Reactor

The Ground Test Reactor is located at General Dynamics, Fort Worth, Texas. The reactor is owned by the United States

TABLE 4-III

Spectra of Several Nuclear Sources (based on W.N. McElroy et al., Nucl. Sci. Eng., November 1964)

Neutron source	A 10 keV to 18 MeV	B 1 MeV to 18 MeV	A/B
SNAP 8	0.68	0.47	1.45
Godiva	0.9998	0.508	1.97
Fission*	0.9995	0.690	1.45
Ground Test Reactor	0.649	0.315	2.06
1/E plus fission	0.468	0.209	2.24
Ford Nuclear Reactor	0.814	0.414	1.97

*Fission $\phi(E) = \left[0.453 \exp(-E/0.965) \sinh(2.29E)^{\frac{1}{2}} \right]$ [3].

Air Force and used primarily for radiation effects work on research and development programs of the United States Government.

Experiments may be arranged providing exposure to steady-state fluxes of neutron and gamma radiation in a number of different chamber configurations. These configurations are: (1) A remotely operated shuttle system with inner dimensions of $2 \times 2 \times 2\frac{1}{2}$ ft; 1.9×10^{12} neutrons (<0.48 eV)/cm²-sec, 2.4×10^{11} neutrons (>2.9 MeV)/cm²-sec, 10^7 rads (C)/sec. (2) Pneumatically operated tubes either 6 in. long and 1.75 in. in diameter or 2.5×0.625 in.; 2.3×10^{12} and 1×10^{12} neutrons (<0.48 eV)/cm²-sec, 2.8×10^{11} and 1.3×10^{11} neutrons (>2.9 MeV)/cm²-sec, 10^7 rads (C)/sec. (3) A vacuum system capable of maintaining a pressure of 1.2×10^{-7} mm Hg, with an available volume 37 in. deep and 20 in. in diameter; 1.2×10^{11} neutrons (<0.48 eV)/cm²-sec, 1.5×10^{11} neutrons (>2.9 MeV)/cm²-sec, 9×10^6 rads (C)/sec. Other experiment chambers provide facilities for testing while the experiment is at cryogenic temperatures.

Table 4-IV presents the ratio of the GTR's neutron flux at various energies (with boral shielding).

TABLE 4-IV

Ratio of the Integrated Neutron Flux of the Ground Test Reactor at Various Energies (with Boral Shielding)

Flux ratio	Energy range	Flux range (neutrons/cm^2-sec-W)	Flux ratio*
$\frac{\phi \text{ th}}{\phi \text{ epi}}$	0-0.48 eV 0.48 eV-∞	5.8 (2) 1.6 (5)	3.6 (-3)
$\frac{\phi \text{ epi}}{\phi \text{ Pu}}$	0.48 eV-∞ 1 keV-∞	1.6 (5) 1.3 (5)	1.2
$\frac{\phi \text{ Pu}}{\phi \text{ Np}}$	1 keV-∞ 0.75 MeV-∞	1.3 (5) 7.2 (4)	1.8
$\frac{\phi \text{ Pu}}{\phi \text{ U}}$	1 keV-∞ 1.5 MeV-∞	1.3 (5) 4.8 (4)	2.7
$\frac{\phi \text{ Pu}}{\phi \text{ S}}$	1 keV-∞ 2.9 MeV-∞	1.3 (5) 2.9 (4)	4.5
$\frac{\phi \text{ P}}{\phi \text{ S}}$	2.4 MeV-∞ 2.9 MeV-∞	3.8 (4) 2.9 (4)	1.3
$\frac{\phi \text{ U}}{\phi \text{ S}}$	1.5 MeV-∞ 2.9 MeV-∞	4.8 (4) 2.9 (4)	1.7
$\frac{\phi \text{ Np}}{\phi \text{ S}}$	0.75 MeV-∞ 2.9 MeV-∞	7.2 (4) 2.9 (4)	2.5
$\frac{\phi \text{ Np}}{\phi \text{ U}}$	0.75 MeV-∞ 1.5 MeV-∞	7.2 (4) 4.8 (4)	1.5
$\frac{\phi \text{ S}}{\phi \text{ Mg}}$	2.9 MeV-∞ 6.3 MeV-∞	2.9 (4) 1.7 (3)	17
$\frac{\phi \text{ S}}{\phi \text{ Al}}$	2.9 MeV-∞ 8.1 MeV-∞	2.9 (4) 1.1 (3)	26
$\frac{\phi \text{ S}}{\phi \text{ th}}$	2.9 MeV-∞ 0-0.48 eV	2.9 (4) 5.8 (2)	50

*To determine actual flux difference, reduce the flux ratio by one, and multiply it by the denominator of the quotient.

OPERATING DESCRIPTIONS OF VARIOUS PULSED RADIATION FACILITIES

KEWB* (Liquid-Fueled Homogeneous Core-Reflected Pulsed Reactor)

KEWB *A* (Spherical Core). The reactor core vessel is spherical and made of stainless steel. It has an inside diameter of 12.3 in. with a maximum wall thickness of 0.22 in. and volume of 13.9 liters. Fuel solution is distilled water of uranyl sulfate enriched to 90% U^{235}. A critical mass of U^{235} is 1680 g for normal core (85% full) and 1150 g for full core. Radiolytic H_2 and O_2 gases are produced during pulsed operation.

KEWB *B* (Cylindrical Core). The core container is a cylinder 3 ft high, 1 ft in diameter and with a volume of 20 liters. Two large concentric control elements with a total diameter of approximately 3 in. enter a vertical thimble in the reactor core. The outer rod is actuated by a high-speed pneumatic device capable of releasing the required reactivity in about 10 msec. The inner rod is the control shutdown rod which may be accurately positioned for control purposes. This is attached to the device mechanism by means of a electromagnet which is deenergized in case of power failure, abnormal indications from scram units, opening of safety interlocks systems, etc., allowing the rod to drop into the solution, providing shutdown. The position is controlled by a reversible electric motor through a gear device and is indicated on the control console.

Pulsed Operation for Both *A* and *B* Cores. A poison rod is removed from the core and the control-shutdown rod is adjusted to the point at which the chain reaction is just self-sustaining at a very low, essentially zero, power level (i.e., delayed critical condition). The poison rod is then inserted and the main control rods, in the case of the KEWB *A* core, or the control-shutdown rod itself, in the case of the KEWB *B* core, are removed to a level which corresponds with the desired excess reactivity input upon removal of the poison rod. The pnuematic removal apparatus is then activated and rapid ejection of the poison rod follows. The reactor is now super-critical and the pulse is allowed to proceed.

*KEWB = Kinetic Experiments on Water Boilers.

There are two inherent shutdown mechanisms that prevent the continuous rise of power level, the gaseous void formation, and the thermal expansion of the fuel. The void and temperature coefficients are both negative over all ranges of the KEWB reactors. The increase in fuel temperature reduces the fuel solution density, limiting the attainable fission rate. The formation of gaseous voids in the fuel solution from the decomposition of water is responsible for the shutdown mechanism, especially at short reactor periods when the decomposition rate is large.

KEWB Reactor [Atomics International (AI) Division of North American Aviation]: The KEWB reactor is located at Conoga Park, California, and was developed by AI. The KEWB pulse width is in the millisecond region and, therefore, the integrated neutron doses are high.

Nuclear Characteristics.

KEWB *B* Core:

1. Prompt neutron lifetime 3×10^{-5} sec.
2. Neutron pulse half-width, 3.0 msec.
3. Peak neutron flux:
     fast ($E > 10$ keV), 2.4×10^{16} neutrons/cm^2-sec;
     resonance flux (0.625 eV $< E <$ 10 keV), 3.0×10^{15} neutrons/cm^2-sec;
     thermal flux ($E <$ 0.625 eV), 7.4×10^{14} neutrons/cm^2-sec.
4. Integrated neutron flux:
     fast, 1.0×10^{14} neutrons/cm^2;
     resonance, 1.3×10^{13} neutrons/cm^2;
     thermal, 3.1×10^{12} neutrons/cm^2.
5. Peak gamma flux, 3.2×10^7 rads (C)/sec.
6. Total gamma dose, 1.4×10^5 rads (C).
7. Neutron energy spectrum, see Fig. 4-2.

KEWB *A* Core:

1. Prompt neutron lifetime, 1.2×10^{-4} sec.
2. Neutron pulse half-width, 10 msec.
3. Neutron energy spectrum, as for *B* core.

TRIGA (Solid-Fueled Heterogeneous Core-Reflected Pulsed Reactor)

Mark I; First Prototype. This facility is located at the General Atomic Division, San Diego, California. The core consists of a cylindrical array of fuel-moderator elements and dummy (graphite) elements. For Mark I, the core is located in a water tank. The fuel-moderator elements are 28.44 in. long and 1.47 in. in outside diameter. The fuel region is 14 in. long, consisting of a solid homogeneous mixture of hydrided uranium–zirconium alloy (8 wt. % uranium enriched to 20% U^{235}). A 4 in.-long graphite reflector, a burnable poison contained in an aluminum wafer, and an aluminum end fixture are inserted at each end of this active portion. The four control rods are made of boron carbide. Two are shim safety rods, one is a regulating rod, and one is used as a transient or pulsing rod. The reflector surrounding the core is a ring-shaped block of graphite 12 in. thick radially, with an 18-in. inside diameter, and 22 in. in height, encased in an aluminum can to prevent water leakage.

Nuclear characteristics are as follows:

1. Prompt neutron lifetime, 6.5×10^{-5} sec.
2. Neutron pulse half-width, 15 msec.
3. Peak neutron flux obtainable (in E-ring fuel-element cavity): fast ($E > 10$ keV), 2.0×10^{16} neutrons/cm^2-sec; thermal, 1.3×10^{16} neutrons/cm^2-sec.
4. Fast neutron dose (> 10 keV), 4.0×10^{14} neutrons/cm^2.
5. Gamma dose rate, 4.7×10^{7} rads (C)/sec.
6. Gamma dose, 9.5×10^{5} rads (C).
7. Neutron energy spectrum, similar to fission energy spectrum in Fig. 4-2.

Mark II. The basic design is the same as for Mark I.

Mark F. This facility has the same basic fuel elements as Mark I but is especially designed for pulsed operation. The core is located near the bottom of a water-filled tank (swimming pool) 14 ft long, 13 ft wide, and 20 ft deep.

Nuclear characteristics are as follows:

1. Prompt neutron lifetime, 4.5×10^{-5} sec.
2. Maximum neutron pulse half-width, 10 msec.

3. Peak neutron flux (F-ring): fast ($E > 10$ keV), 3.4×10^{16} neutrons/cm^2-sec; thermal, 2.4×10^{16} neutrons/cm^2-sec.
4. Fast neutron dose (> 10 keV), 4.1×10^{14} neutrons/cm^2.
5. Gamma dose rate, $> 5.0 \times 10^7$ rads (C)/sec.
6. Gamma dose, 1.0×10^6 rads (C).
7. Neutron/gamma ratio, 0.9.
8. Neutron spectrum, see Fig. 4-2.

Pulsed Operation of All TRIGA Reactors. Reactivity insertion is accomplished by removal of control rods containing poison material. The reactor is first brought to delayed critical operating condition utilizing a 5 C polonium–beryllium start-up source. The initial position of the drive piston is adjusted so that the desired amount of reactivity will be inserted upon removal of the transient rod from the core. The transient rod is removed pneumatically in about 0.75 sec. Reactor power then rises, falling to a steady-state level as the zirconium hydride moderator is directly headed by interaction with the fission fragments, causing a shift of the thermal-neutron spectrum, thereby resulting in a larger loss of neutrons in the water channels.

Several TRIGA reactors are in operation in addition to the TRIGA facility at General Atomic, Torre Pines, California. These reactors are located at Diamond Ordnance Fuze Laboratory, Washington, D.C.; Armed Forces Radiobiology Research Institute, Bethesda, Maryland; and Norair Division of Northrop, Hawthorne, California. The TRIGA reactor produces high integrated neutron doses. Because of its millisecond pulse width it is useful for testing of electronic piece parts and equipment in a pulsed environment.

Bare Critical Assemblies

Godiva II. The major section of the core assembly is a right circular cylinder 7 in. in diameter, $5\frac{1}{2}$ in. high, with a spherically shaped top section. A coaxial cylindrical cavity in the bottom of the main section accommodates the safety block. Two control rods and a burst rod also enter the main section from the bottom. Screws run by electric motors drive the control rods and pneumatic cylinders drive the safety blocks, burst rod, and a polonium–beryllium neutron source used to initiate the chain reaction in the assembly. Total core mass is 57,700 g of 93.2% enriched uranium. Operation is entirely by remote control from a building $\frac{1}{4}$ mile away.

Nuclear characteristics are as follows:

1. Prompt neutron lifetime, 6×10^{-9} sec.
2. Maximum neutron pulse half-width, $80 \, \mu$sec.
3. Total neutron release, 1.4×10^{16} neutrons.
4. Peak neutron flux ($E > 1$ keV), 2×10^{17} neutrons/cm^2-sec.
5. Peak gamma dose rate, 10^7 rads (H_2O)/sec.
6. Neutron energy spectrum, see Fig. 4-3.

Pulsed operation is as follows: First the assembly is brought to delayed critical configuration by a short, low-level power run. This is done by raising the polonium–beryllium neutron source close to the assembly, raising the safety block to the fully inserted position, and inserting the control rods in sequence and adjusting for delayed critical operation.

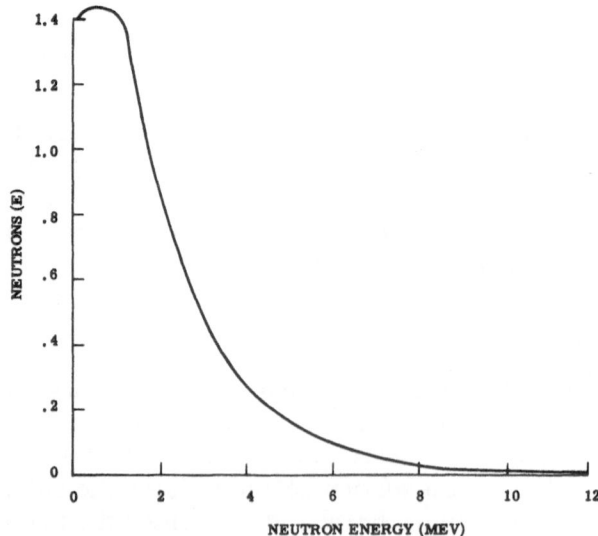

Fig. 4-2. Energy spectrum of fission neutrons.

The source is then retracted and the safety block is retracted to allow the neutron population to decay. The safety block is then inserted rapidly and the burst rod is fixed into the assembly and a super-prompt-critical configuration is achieved. The resulting power excursion is automatically stopped by thermal expansion, bringing the assembly below prompt critical to a steady-state power level of several kilowatts. The burst rod and safety block are then retracted rapidly, bringing the assembly below critical.

KUKLA. The assembly is in the shape of a sphere $6\frac{3}{4}$ in. in diameter. The main sphere control rods and safety blocks are made of cast oralloy enriched to 93.2% U^{235} and the burst rod is made of 4% molybdenum and 96% normal-enriched oralloy. All oralloy surfaces are nickel-flashed to prevent oxidation of its oralloy. The main sphere consists of four sections, and entering the sphere from the bottom are the safety blocks, two control rods, and a burst rod.

Nuclear characteristics are as follows:

1. Total fission yield, 2×10^{16} fissions.
2. Maximum burst half-width, $60\,\mu$ sec.
3. Total radiation dose per burst 1 m away: gamma, 60 rads (C); neutron, 7400 rads (C).

Other characteristics are the same as for Godiva II. Operation is the same as for Godiva II.

SPRF (Sandia Pulse Reactor Facility). The reactor is modeled after Godiva II. It is mounted on a four-legged instead of a three-legged (as Godiva II) structure. The SPRF, located at the Sandia Corporation, Albuquerque, New Mexico, is a reactor similar to the Godiva II which was used for radiation effects experiments at the Los Alamos Scientific Laboratory. In fact, the SPRF reactor is identical to the Godiva II except that increased reactivity is available, and additional safety factors have been added.

Nuclear characteristics are as follows:

1. Maximum neutron pulse half-width, $50\,\mu$ sec.
2. Peak neutron flux, 3×10^{17} neutrons/cm^2-sec.
3. Neutron dose (>1 keV), 2.3×10^{13} neutrons/cm^2.
4. Peak gamma dose rate, $2-6 \times 10^7$ rads (H_2O)/sec.

TABLE 4-V

Summary of Nuclear Characteristics of Pulsed Reactors*

Characteristic	KEWB A (Max.)	KEWB B (Max.)	TRIGA Mark I Mark II (E-ring)	TRIGA Mark F (E-ring)	Godiva II (Max.)	KUKLA (Max.)	SPRF (Max.)	ORNL Super Godiva (Max.)
1. Prompt neutron lifetime (sec)	1.2×10^{-4}	3×10^{-5}	6.5×10^{-5}	4.5×10^{-5}	6×10^{-9}	10^{-8}	6×10^{-9}	—
2. Fast neutron leakage per pulse (neutrons)					1.4×10^{16}	1.4×10^{16}	2.8×10^{16}	1.3×10^{17}
3. Fast neutron flux available (neutrons/ cm²-sec)		2.4×10^{16} (>10 keV)	2.0×10^{16} (>10 keV)	3.4×10^{16} (>10 keV)	2.3×10^{17} (>1 keV)	2.0×10^{17} (>1 keV)	3×10^{17} (>1 keV)	3.2×10^{18} (>1 keV)
4. Maximum integrated flux (neutrons/cm²)		1×10^{14} (>10 keV)	4×10^{14} (>10 keV)	4.1×10^{14} (>10 keV)	1.6×10^{15} (>1 keV)	1.4×10^{15} (>1 keV)	2.3×10^{15} (>1 keV)	7.2×10^{15} (>1 keV)
5. Peak gamma flux per pulse (rads/sec)		3.2×10^{7} (C)	4.7×10^{7} (C)	$> 5.0 \times 10^{7}$ (C)	10^{7} (H₂O)	2×10^{7}	6×10^{7} (H₂O)	—
6. Total gamma dose per pulse (rads)		1.4×10^{5} (C)	9.5×10^{5} (C)	1.0×10^{6} (C)	10^{3}	2×10^{3}	5×10^{3} (H₂O)	4.0×10^{8}
7. Reactor period (μsec)	3000	1000	5000	4000	20	17	20	13
8. Neutron pulse half- width (μsec)	10,000	3000	15,000	10,000	80	60	50	38

*Supplementary information can be found in Radiation Effects Information Center Report No. 3, Part II, Battelle Memorial Institute (Columbus, Ohio), September 15, 1963.

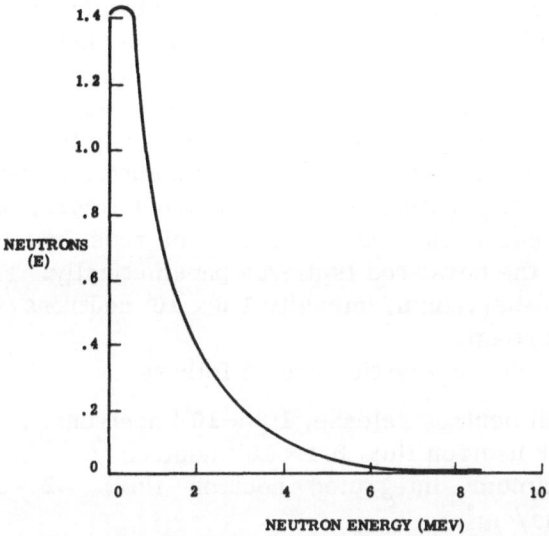

Fig. 4-3. Neutron energy spectrum of the SPRF (Sandia Pulsed Reactor Facility).

5. Gamma dose, $2-5 \times 10^3$ rads (H_2O).
6. Neutron spectrum, see Fig. 4-3.

Operation is essentially the same as for Godiva II.

ORNL* Super Godiva.   The reactor core is a right circular cylinder $7\frac{7}{8}$ in. in diameter and $7\frac{1}{16}$ in. long. The bottom is rounded so that a more uniform neutron flux distribution is obtained below and to the sides of the reactor. The cylinder consists of two shells, one fitting inside the other. The other shell is $2\frac{1}{2}$ in. thick, consisting of a stack of annular disks made of an alloy of 90% tungsten-10% molybdenum and held together by nine $\frac{7}{8}$-in.-diameter bolts made of the same material, which thread into the bottom dish. It is supported from the reactor structure by three steel mounting fingers attached to the top, permitting unrestrained thermal expansion in all directions.

The inner shell is $2\frac{3}{4}$ in. in diameter, $\frac{1}{2}$ in. thick, $7\frac{1}{4}$ in. long, and has a central steel plug $1\frac{7}{8}$ in. in diameter which is used as a power flattening device to reduce the maximum

*ORNL = Oak Ridge National Laboratory.

temperature in the safety block, in order to allow free thermal expansion of the safety block in all directions. The safety block is supported by a steel shaft which actuates it vertically. The block is pneumatically driven and can be inserted in 40 msec. A spring inside the cylinder expels the safety block in the event of loss of air pressure or electrical power.

Three rods penetrate the outer shell of the core, two control rods, and one burst rod. The control rods are electrically driven and the burst rod is driven pneumatically. The source is polonium–beryllium, intensity 1.6×10^6 neutrons/sec, and is used for start-up.

Nuclear characteristics are as follows:

1. Total neutron release, 1.3×10^{17} neutrons.
2. Peak neutron flux, 3.2×10^{18} neutrons/cm^2-sec.
3. Maximum integrated neutron flux, 7.2×10^{13} neutrons/cm^2.
4. Maximum neutron pulse half-width, 38 μsec.

Operation is similar to Godiva II, KUKLA, and SPRF. Table 4-V summarizes the nuclear characteristics of pulsed reactors.

FACILITIES FOR TRANSIENT IONIZATION EFFECTS TESTING

Particle generators in general can take any number of forms. A complete listing of these and their availability for radiation effects testing may be found in the Battelle Memorial Institute Report No. 31, Part II [1]. The particle generators can be one of the following types: Linear Accelerators, Flash X-ray, Betatrons, fixed-frequency Cyclotrons, frequency-modulated Cyclotrons, variable-frequency Cyclotrons, Van de Graaff, Dynamitron, Fexitron, I.C.T., Cockcroft–Walton, Statitron, and Synchrotron.

Of these, the Linear Accelerators are most popular for electronic radiation effects studies. High-energy Flash X-ray machines are now available at Physics International, and these are good because of their ability to irradiate rather large volumes with an approximately constant ionizing dose rate.

The Van de Graaff machines belong to the family of "potential drop machines," meaning that the design is based on letting the desired particle gain the required energy by falling down a potential hill. The Van de Graaff machines and the other machines of this category, such as the Dynamitron, Fexitron, and Cockcroft—Walton, are mainly used to provide mono-energetic particles, either protons or neutrons. These machines are therefore particularly suited for low-dose cosmic radiation effects studies on electronics.

Linear Accelerators

The Linac (Linear Accelerator) generates high-energy electrons, which are either used directly for irradiation of relatively thin samples (for instance, direct exposure of individual transistors in a circuit) or are stopped to generate by the bremsstrahlung-process X-radiation. The important consideration is to assure uniform release of energy through the sample. This can be accomplished by 10 MeV electrons in $9/10$-in. silicon, for instance. A high-energy Linear Accelerator is also sometimes used to generate fission spectrum neutron beams by bombarding a sample of U^{238} with the electron beam.

The General Atomics 45 MeV Linear Accelerator (Linac) is located at the General Atomic Division, San Diego, California. Maximum beam current is 700 mA, beam energy is 43 MeV, pulse length is 0.5—15 μsec (adjustable). The maximum average neutron yield is 3×10^{16} neutrons/cm^2-sec. Maximum dose rate over a small area is 10^{10} rads (C)/sec. Operating cost is about $156.00 per hour. Availability is two weeks.

Several Linacs are available for short-pulse-width type radiation effects experiments. These Linacs are located at Rensselaer Polytechnic Institute, Troy, New York (45 MeV); Hughes Aircraft, Fullerton, California (10 MeV); General Atomic, Torre Pines, California (45 MeV); WSMR (10 MeV); and Boeing, Seattle, Washington (10—25 MeV).

Flash X-Ray Facilities

300 and 600 keV Flash X-ray machines are available. They produce pulse widths from 0.1 to 2 μsec, and have peak rates from 10^6 to 10^8 rads (C)/sec over small sample sizes.

Large Flash X-ray equipment is available at Physics International, Berkeley, California and at Ion Physics Corp., Burlington, Mass. It provides a continuous spectrum of X-rays to 2 MeV. 10^7 to 10^{11} rads (C)/sec peak rates are produced for pulse widths of 20 to 40 nsec.

DOSIMETRY*

The monitoring of magnitude and spectrum of the radiation environment during radiation testing is performed by utilizing radiation dosimetry. The dosimetry is in general provided as a service by the radiation test facility personnel. However, as the accuracy of radiation test result reports depends on knowledge of dosimetry and its accuracy, it is the purpose of the following to briefly describe the various dosimetry methods and their characteristics.

Neutrons

Neutron dosimetry in general is provided by passive detectors in the form of either fission or resonance foils. The foils are materials with well-known and quite accurate threshold activation energies, for instance U^{235}, whose fission threshold is 1.5 keV. The threshold energy is defined as that energy or greater that a neutron must possess to cause fission of the U^{235} sample. Analysis of the sample subsequent to the exposure will reveal how many fissions took place, which is therefore a measure of the amount of neutrons, in the particular radiation beam, that had kinetic energies greater than 1.5 keV. Errors from ± 15 to 30% are encountered with this type of dosimeter.

The resonance foil consists of materials which exhibit very high capture cross sections for neutrons within a very narrow range. Aluminum exhibits such a resonance at 9.1 keV, vanadium at 3.0 keV, copper at 0.57 keV, and manganese at 0.26 keV. The resonance foil is generally covered by cadmium to exclude activation of the foils by thermal neutrons. Other typically used neutron foils are listed in Table 4-VI.

*Supplementary information can be found in B. P. Fairand and E. N. Wyler, Radiation Effects Information Center Report No. 33, Battelle Memorial Institute (Columbus, Ohio), September, 1963.

TABLE 4-VI

Neutron Foils

Material	Reaction	Threshold
Pu^{239}	n, fission	1-10 keV determined by shield
Np^{237}	n, fission	0.75 MeV
U^{238}	n, fission	1.5 MeV
U^{235}	n, fission	1.5 keV
Th^{232}	n, fission	1.75 MeV
S^{32}	(n, p) P^{32}	2.9 MeV
Mg^{24}	(n,p) Na^{24}	6.3 MeV
Al^{27}	(n, α) Na^{24}	8.1 MeV
Au^{197}	(n, γ) Au^{198}	0.3 eV
Co^{59}	(n, γ) Co^{60}	1.32 eV

Then, by using a complement of foils with different threshold energies, the reactor spectrum can be mapped. In most reactors this is done as needed, or at least every six months, by the reactor personnel or by personnel from companies which specialize in the field of dosimetry. E.G. and G. of Boston, Massachusetts is one such company. The spectrum map of the neutron energy is as accurate as the dosimetry, i.e., ±15 to 30% error at those points in the map represented by a specific dosimeter. However, the spectrum can fluctuate widely between such points, and an overall error of the entire spectrum map is ± 40 to 50%.

While testing, it is common practice to include a sulfur dosimetry sample with the experiment. As noted in Table 4-VI it will supply information as to the amount of neutrons with energies greater than 2.9 MeV. This dosimetry point then provides a check of the behavior of the reactor at the particular time of the experiment and can be used to scale the spectrum map accordingly. The actual spectrum of the reactor output is influenced by fuel characteristics and can vary from week to week.

Another convention amongst electronic radiation experimenters is the habit of quoting the > 10 keV to > 3 MeV integrated dose ratio for a particular experiment. It is the ratio of the plutonium foil reading to the sulfur foil reading and provides a measure of the hardness of a particular spectrum. For instance, the TRIGA MK I reactor has a typical 10 keV to 3 MeV ratio of 6 to 10, whereas the spectrum of neutrons generated by bombarding U^{238} with high-energy electrons has a typical 10 keV to 3 MeV ratio of 20 to 25. The TRIGA MK I therefore provides a much harder spectrum, i.e., it has a greater amount of high-energy neutrons in its spectrum.

General Electric has developed a new technique for neutron dosimetry, which is based on the actual counting of neutron-induced physical fission tracks in a low Z material.* The technique is an advance in the neutron dosimetry field because of the fact that the registered dose is a permanent record, as opposed to the foil dosimeters, whose readings decay with time.

The radiation degradation characteristics of transistors can be used as a dosimeter method for neutron doses greater than 10^{12} neutrons/cm^2 but less than about 10^{14} neutrons/cm^2. However, a rather large sample size of the same type of transistor must be tested to provide calibration of the device as a dosimeter, and in general the method is only used as a checking technique on the primary dosimetry.

Gamma and X-Rays

Dose. The dose may be calculated from the measurements of the dose rate and pulse shape by the instruments described below under Dose Rate. The integrated dose can also be measured by passive dosimeters such as the following:

1. Cobalt glass made by Bausch and Lomb changes its ultraviolet transmittance as a function of the absorbed dose in the glass. The dose range covered is from 10^3 to 10^6 rads. Care should be exercised to ensure that the same dose rate is used for calibration as is expected during use. When this

*Patent being applied for.

is done, the error of a cobalt-glass dosimeter is approximately 10%.

2. A totally rate-independent dosimeter is available in the NRL-developed thermoluminescence detector. It consists of lithium fluoride, which is activated with magnesium. Once irradiated it will emit light when heated. The total light energy emitted is a measure of the absorbed dose. It has a range from 10^{-2} to 10^5 rads and an error of about 5%.

3. Silver-activated phosphate glass is being used as a passive dosimeter for the absorbed dose. Its fluorescence properties are changed in proportion to absorbed dose. It is made by Bausch and Lomb in rods 6 mm long and 1 mm in diameter. It is somewhat spectrum-dependent, and in some cases energy shields are used to correct for this. The range is from 10 to 10^4 rads, with an error of about 5%.

4. A calorimeter may be used to advantage in some cases, since generated heat and absorbed dose are directly relatable.

5. Chemical dosimeters are also available. These cover ranges from 10^2 to 10^7 rads, depending on type, and have errors from 2 to 10%.

Dose Rate.   To measure dose rate in a reactor or Linac experiment, standard instruments, such as ionization chambers, semiconductor diodes, and the Semi-Rad (a rate detector described in Chapter 5) are used. Of these, the semiconductor diode or $p-i-n$ detector and the Semi-Rad provide the fastest response and are generally used in Linac and Flash X-ray experimentation.   The ionization chamber is fast enough to follow the relatively slow rise of the pulse in a pulsed reactor and is often used to monitor the reactor performance.

The semiconductor devices are linear for ranges from 10^4 to 10^9 rads (C)/sec. The accuracy of the measurement depends on the data display. For fast pulses, the display is in the form of oscilloscope pictures, which may be read with an error of about ± 5%. For the rates beyond 10^9 rads (C)/sec the Semi-Rad may be used. It does not saturate for rates less than 10^{13} rads (C)/sec and has an inherently very fast response.

The error bands quoted are based on the assumption that careful calibration by well-established techniques has been performed.   (See National Bureau of Standards literature.)

REFERENCE

1. D. J. Hamman and W. H. Veazie, Jr., "Survey of Particle Accelerators," Radiation Effects Information Center Report No. 31, Part II, Battelle Memorial Institute (Columbus, Ohio), September 1963.
2. Saelens, R.G., et al., "Radiation Effects," Electronic Industries (October 1962).

BIBLIOGRAPHY

Fairand, B. P., and E. N. Wyler, "A Survey of Current Research and Development in the Field of Dosimetry," Radiation Effects Information Center Report No. 33, Battelle Memorial Institute (Columbus, Ohio), September 1963.

Hamman, D. J., and W. H. Veazie, Jr., "Survey of Irradiation Facilities," Radiation Effects Information Center Report No. 31, Part I, Battelle Memorial Institute (Columbus, Ohio), September 1963.

Hamman, D. J., and W. H. Veazie, Jr., "Survey of Particle Accelerators," Radiation Effects Information Center Report No. 31, Part II, Battelle Memorial Institute (Columbus, Ohio), September 1963.

Hine, G. J., and G. L. Brownell, Radiation Dosimetry, Academic Press Inc. (New York).

Jones, D. C., et al., "Radiation Effects Experimental Procedures," Radiation Effects Information Center Report No. 35, Battelle Memorial Institute (Columbus, Ohio), August 1964.

McElroy, W. N., et al., "Advances in the Quantitative Aspects of the Activation Method of Neutron Flux-Spectral Determination for Radiation Effect Studies," Nucl. Sci. Eng. (November 1964).

Soodak, H., Reactor Handbook, Vol. III, Interscience Publishers, Inc. (New York), 1962.

Whyte, G. N., "Principles of Radiation Dosimetry," John Wiley & Sons, Inc. (New York), 1963.

CHAPTER 5

The Nuclear Instrumentation System

INTRODUCTION

In the previous chapters the reader has been presented with essential information for the task of designing electronic equipment which must perform in a nuclear environment. It is the purpose of the last two chapters to present examples of possible application of this knowledge. It should be emphasized that it is not the purpose of the following to dictate fast and unyielding rules for the design of nuclear-hard electronic equipment. Rather, the intent is to present the reader with guidelines which may stimulate creative and unique design approaches. The design of nuclear-hard electronic equipment is a new technology with plenty of room for the application of new design techniques.

This chapter will introduce the reader to the field of nuclear instrumentation, i.e., that portion of the overall instrumentation system which is concerned with the sensing of various types of nuclear radiation, as well as electronic techniques which directly exploit the characteristics of nuclear radiation. The chapter is divided into the following sections:

1. Uses for Nuclear Instruments. This section treats the nuclear environments that can be sensed and gives the reader an insight into the environmental characteristics.
2. Nuclear Instrumentation Designs. Examples of nuclear instrumentation are given in this section. The reader may find occasion to have a specific application for these.

3. System Design — Missions in Space. Examples of entire instrumentation systems are described in this section.

USES FOR NUCLEAR INSTRUMENTS

The uses for nuclear instruments can be listed as follows:

1. Cosmology, the measurements of nuclear radiation in the space and earth environment.
2. Nuclear and particle physics, the continuing investigation directed toward understanding the makeup of matter, the atom, and the nucleus.
3. Radiochemistry, the utilization of nuclear radiation to alter organic materials.
4. Radiology, the utilization of nuclear radiation in the medical field.
5. Industrial applications, where the unique properties of nuclear radiation are exploited in various ways.
6. Nuclear radiation detection, the use of nuclear instruments to detect and report the presence of nuclear radiation.

Of these, (1) and (2) are covered extensively in the following, as it is in these areas that many of the nuclear instruments used in the other fields were originally designed and developed.

Cosmology

Cosmology is defined as the activity associated with the investigation into the origin of the universe and is concerned with propagation of nuclear radiation from outer space. The investigations center on the search for the following types of nuclear radiation: galactic cosmic rays, solar cosmic rays, high-energy gamma rays, low-energy gamma rays, Van Allen belt particles, interplanetary plasma, and radioactive emissions from unstable element isotopes in the interplanetary medium (used in the search for elements in the solar system). A brief description of each of these follows below to acquaint the reader with the environment.

Galactic Cosmic Rays. As described in Chapter 1, cosmic rays are positively charged heavy particles, generally consisting of helium nuclei. Galactic rays are those originating outside of the solar system. The detection of higher-energy galactic cosmic rays has taken place on the earth's surface during the last decades. However, it is not possible to tell the origin of these charged particles, as they have interacted many times with magnetic fields en route. Because lower-energy cosmic rays are partly reflected by the atmosphere and because it is desirable to study this cosmic ray albedo, observations must be made with a nuclear instrument aboard an orbiting space vehicle.

Solar Cosmic Rays. These are high-energy particles originating from large solar flares. Up to 10^5 protons/cm^2-sec with energies greater than 50 MeV can be produced by the largest solar flares. The study of these solar cosmic rays and their time characteristics produces information about the nature and makeup of the interplanetary medium. The rapid rise in intensity of these particles would indicate a direct magnetic path from sun to earth, whereas at other times a slow rise in intensity indicates a reflection of the particles to earth from other local magnetic fields within the interplanetary system. That solar flares are sometimes never detected in the earth's environment was evidenced by the fact that the Mariner IV Mars Probe detected a number of solar flares on its journey through space. Two of these were not detected by detection systems on earth.

High-Energy Gamma Rays. Gamma rays of high energy, when detected in the interplanetary medium and in the upper fringes of the earth's atmosphere, are important because they provide a direct message as to where in the universe they originated. Since they are electromagnetic waves, they are not influenced by magnetic fields and, unless annihilated through interactions with particles, they will propagate on a straight path from their source to the detector. Because of this fact an early orbiting satellite was equipped with a special gamma detector to measure gamma rays with energies greater than 50 MeV. The experiment was designed by Kraushaar and Clark of MIT and is more fully described below.

Low-Energy Gamma Rays. In the terminology of cosmology, these are gamma rays with energies from 1 keV to 10 MeV.

They are mainly produced by the sun, but are also constituents in the radiation arriving from outside of the galaxy.

Van Allen Belt Particles. These are particles, electrons and protons, trapped in the earth's magnetic field. There are two suggested sources for these: (1) particles injected by solar flares and (2) neutrons, produced as an albedo from cosmic rays, that decay into protons and electrons, one of each resulting from one neutron. These particles are subsequently captured by the magnetic field. Many orbiting satellites have carried nuclear instruments to map the trapped particle belts. Some of these instruments are described later.

Interplanetary Plasma. This plasma is very complex. As described in Chapter 1, it originates near the sun and can in confined regions reach densities of 100 ions/cm^3. The study of this plasma will continue to take nuclear instruments into the solar space, preferably at high solar latitudes. Mariner II on its voyage to Venus provided excellent data about this plasma.

Search for Elements. The tools used in the search for elements in the solar system consist of mass spectrometers which can identify the element by the emission of gamma rays from radioactive isotopes. The most accurate method consists of collecting or trapping a sample of the element, whether gas, as in the interplanetary medium, or a solid, as in the case of actual contact with a planet, a meteoroid, or an astroid. The sample is made radioactive and the resulting emissions noted and used to identify the element.

The nuclear instruments utilized to measure the environments just described consist of one or a combination of the following: (1) semiconductor detectors, (2) Geiger–Müller counters, (3) proportional counter arrays, (4) ionization chambers, and (5) scintillation detectors. These are carried by satellites or space probes, whose trajectories are well calculated in order to keep a close track of which portions of near-earth or the interplanetary system are traversed. The first Geiger–Müller tube was sent into the Van Allen belt aboard a captured German V-2 rocket in the 1940's, marking the beginning of an ever-accelerating advance toward the understanding of the origin of the universe and the solar system.

Nuclear and Particle Physics

Nuclear radiation research utilizes basically the same nuclear instruments as described above. Nuclear instruments unique to this field are: (1) the spark chamber, (2) the fast neutron spectrometer [13], and (3) semiconductor particle detectors. The spark chamber and the semiconductor particle detector are described in the following section.

NUCLEAR INSTRUMENTATION DESIGNS

The increased utilization of nuclear instruments aboard satellites and space probes in the near-earth and interplanetary space has produced a number of quite unique instrumentation designs. In the following, a few of these will be described in detail in order to acquaint the reader with this new field of nucleonics design. Since these nuclear instruments are to be flown aboard space vehicles, they must operate correctly and survive the shock and vibrations associated with the boost phase of the space mission. There are also rigid requirements for performance reliability during exposure to space radiation, temperature, and vacuum. The designing of an electronic package which will meet these requirements is no simple task and requires the services of an experienced packaging engineer. Typical requirements for nuclear instruments are shown in Table 5-I. Radiation requirements are specified if prolonged flight in the Van Allen belts is expected or if operation will occur in the vicinity of a nuclear reaction, as, for instance, aboard a nuclear rocket propelled vehicle.

TABLE 5-I

Typical Requirements for Nuclear Instruments Aboard Space Vehicles

Condition	Requirement
Temperature	−10 to 70°C
Vibration	0.1 g^2/cps from 30 to 2000 cps
Acceleration	10 g's for 2.5 min
Shock	3 50-g impulses (half sine) of 2 msec duration in forward direction
Vacuum	4 days at +40°C and 10^{-5} mm Hg

Spectrometers

The particles and electromagnetic waves encountered in near-earth and solar space have a continuous energy spectrum. It is important for researchers to obtain data on the distribution of the radiation with respect to energy. It is to gain these data that a variety of spectrometers have been sent into space aboard satellites and space probes. Three types are described below: (1) electron spectrometer, (2) cosmic ray spectrometer, and (3) X-ray spectrometer.

Electron Spectrometer. A low-energy electron spectrometer was designed by Lockheed Missiles and Space Company [1] for a polar orbit satellite in an effort to measure the energy spectrum of auroral particles. Figure 5-1 shows the instrument in cross section.

The electron energy range of interest was from 22 to 113 keV. The particles enter the Plexiglas collimator in the lower left hand corner of the instrument. The use of Plexiglas reduces the possibility of having scattered electrons enter the main chamber of the instrument, since Plexiglas, as pointed out in Chapter 2, is a good electron absorber. Once inside the main chamber of the instrument the charged particles come under the influence of a known magnetic field generated by the ceramic magnet. The magnet surfaces are also covered by Plexiglas to reduce scattering. The electrons in the magnetic field are deflected by an amount which is in proportion to their energy, and a spectrum determination is therefore accomplished. The deflected electrons finally enter one of four

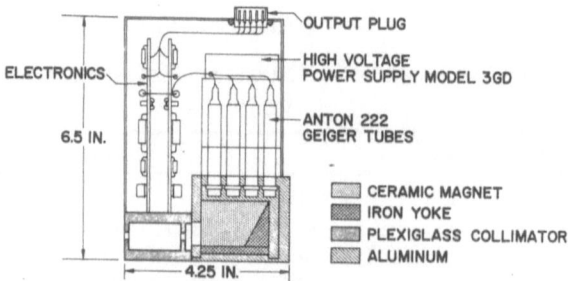

Fig. 5-1. Electron spectrometer (reproduced with the permission of IEEE, from J. B. Reagan and R. V. Smith, IEEE, NS-10, January 1963).

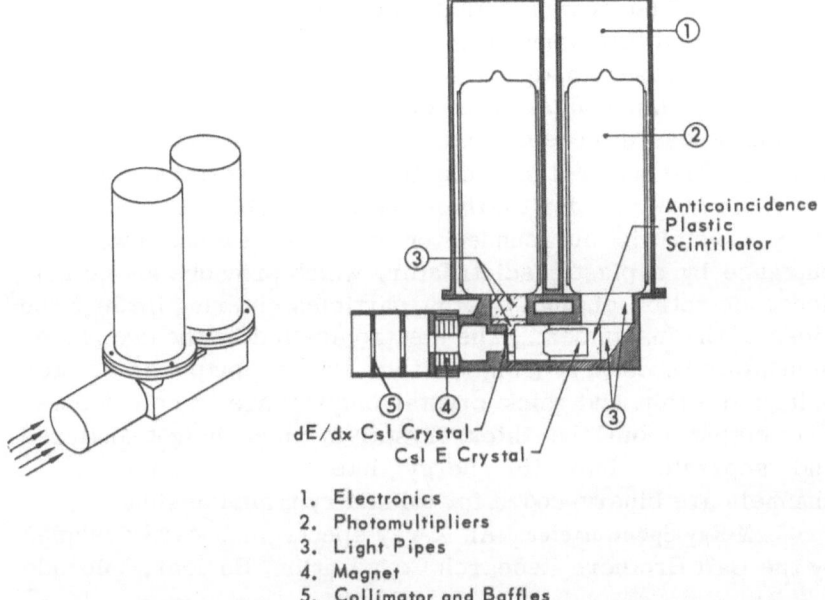

1. Electronics
2. Photomultipliers
3. Light Pipes
4. Magnet
5. Collimator and Baffles

Fig. 5-2. Cosmic ray spectrometer (reproduced with the permission of IEEE, from J.H. Rowland et al., IEEE, NS-10, January 1963).

Geiger–Müller counter tubes, two of which are covered with thin aluminum shields to prevent scattered low-energy electrons from entering. The tube walls are covered with lead shielding to prevent entrance of particles from the sides. The outputs from each of the four G–M tubes were integrated and had a range from 10 to 10^4 counts/sec. 10 counts/sec was equal to 5 V DC and 10^4 counts/sec equal to 2.5 V DC. A radioactive source was included in the assembly as an inflight calibrator. The four outputs were commutated continuously. Two such units were flown, consuming 2.25 W at 28 V DC.

Cosmic Ray Spectrometer. This instrument, also designed by Lockheed Missiles and Space Company [2], was built to measure fluxes and spectra of protons in the range from 10 to 100 MeV. The spectrometer is shown in cross section in Fig. 5-2. The collimator is shown in the lower left-hand side of the instrument. It contains 15 g/cm^2 of brass shielding in all directions except that desired for entrance. Once they have entered, the charged proton particles are further collimated by a magnet before they encounter a thin CsI crystal. The thickness of about 0.01 in. is chosen for this crystal so that a 95-MeV proton will lose 0.5 to 0.7 MeV of energy passing through it. Behind the thin crystal in the entrance tube is yet another CsI crystal whose dimensions are chosen to just stop a 100-MeV proton. The output threshold for the thin crystal was set for about 0.5 MeV, and for the thick crystal for about 10 MeV. The thick crystal is surrounded on all sides except toward the entrance by a plastic scintillator, which provides anticoincidence detection of high-energy particles entering through the sides of the instrument. The electronics fed by the two photomultiplier tubes are designed such that no output is created unless the thin and thick crystal outputs are in coincidence. The output from the thick crystal is pulse-height analyzed and separated into 16 energy intervals. Each of the 16 channels are binary-coded for telemetry transmission.

X-Ray Spectrometer. An X-ray spectrometer was designed by the Ball Brothers Research Corporation, Boulder, Colorado

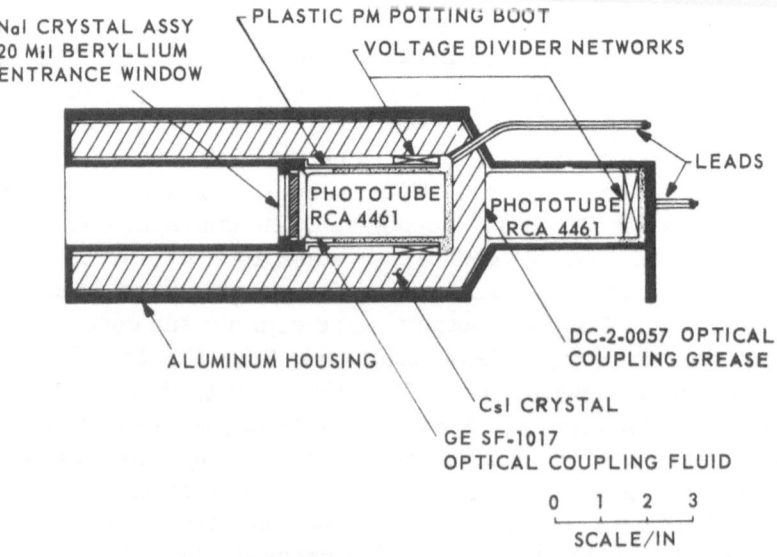

Fig. 5-3. X-ray spectrometer (reproduced with the permission of IEEE, from D. B. Hicks, L. Ried, Jr., and L. E. Peterson, IEEE, NS-12, February 1965).

[3]. The spectrometer was flown on NASA S-57 Orbiting Solar Observatory-C, and was designed to measure the intensity, spectrum, and time variations of solar X-rays. The spectral range was made to cover photons from 7 to 190 keV. The detector shown in cross section in Fig. 5-3 consists of a central NaI (Tl) crystal surrounded by an anticoincidence shield of CsI (Tl), which opens toward the desired entrance direction. The acceptance angle of the detector opening is 25.5°. Both the NaI scintillator and the CsI scintillator are viewed by an RCA 4461 photomultiplier tube. The NaI crystal is 3.5 cm in diameter and 0.5 cm thick, and has an entrance window of beryllium 0.020 mils thick. NaI was selected because of its relatively high photoelectric absorption coefficient up to about 300 keV, combined with a high light output per unit energy loss.

The CsI sleeve surrounding the NaI detector crystal acts as an active shield in the following manner: (1) It prevents low-energy photons from entering from the sides due to its thickness. (2) Experiments with Cs^{137} Co^{57} radioactive sources permitted an anticoincidence discriminator to be adjusted to 85 keV, which during flight would signal that photons with energies greater than 85 keV had entered from the side. (3) Any charged particles (cosmic rays) entering the detector would be detected by an anticoincidence discriminator set for a threshold of 3.2 MeV. This discriminator level was determined from the fact that minimum ionizing particles have a loss of energy in the CsI crystal to trigger the 3.2 MeV discriminator after traversing only 0.6 cm in the CsI crystal.

The output from the NaI crystal photomultiplier feeds eight discriminators providing spectrum division into the following ranges: > 7 keV, 7–12 keV, 12–21 keV, 21–36 keV, 36–63 keV, 63–110 keV, 110–190 keV, and > 190 keV. The instrument weighs 23 lb and requires 420 mW of power at 28 V DC. It has met the following qualification tests: (1) sinusoidal vibration up to 20 g's; (2) steady-state acceleration, 18 g's; (3) temperature, −10 to + 40°C for 8 hr periods; and (4) thermal vacuum tests: 8 hr period at −10°C and 10^{-5} mm Hg, 4 day period at + 40°C and 10^{-5} mm Hg. The NaI detector output varied ±4% over the temperature range.

Scintillation Detector

As can be seen from the description of the three spectrometers, the scintillation detector is a widely used nuclear
instrument for spectral analysis. The detector in all cases
consists of a scintillation crystal coupled to a photomultiplier.
The scintillation crystal is mostly NaI(Tl), although CsI and CaI
are also extensively utilized. The selection of the correct crystal cannot be made until the specific application of the instrument is determined.   The scintillator is surrounded by an
appropriate thickness of shielding material which determines
the minimum energy threshold of the detector. As the nuclear
radiation traverses the scintillator, it loses energy by ionization and produces a light burst, whose intensity is proportional
to the energy of the ionizing radiation. The light burst is
registered by the photomultiplier tube and amplified to a
signal level, which is adequate for further amplification by
electronic means.   The fast rise-time pulses from the photomultiplier (PM) are pre-amplified by an amplifier, which with
unity gain serves as an impedance matcher between the PM
and the remainder of the electronics. Its characteristics include the capability for following fast rise-time pulses (≥ 10
nsec) and a wide dynamic range of linearity. The pulse is
subsequently fed to a threshold detector, generally called a
pulse-height discriminator. A number of these may be used at
different threshold levels, such as in the X-ray spectrometer
described above. The discriminator is a binary device, registering a zero output when the radiation energy is too low to
produce the necessary output from the PM, and registering a
1 output when energies higher than the threshold energy are
received by the PM. The output from the discriminator triggers
a monostable multivibrator, whose output in turn feeds an
integrator.   The output of the integrator to conform to IRIG
telemetry standards is adjusted such that the range of expected
counting rates from the multivibrator covers a voltage range
from 0 to 5 V DC.   Figure 5-4 shows a typical calibration
curve.

High-Energy Gamma Ray Telescope

In 1960, W. L. Kraushaar and G. W. Clark of the Massachusetts Institute of Technology, to aid their research on

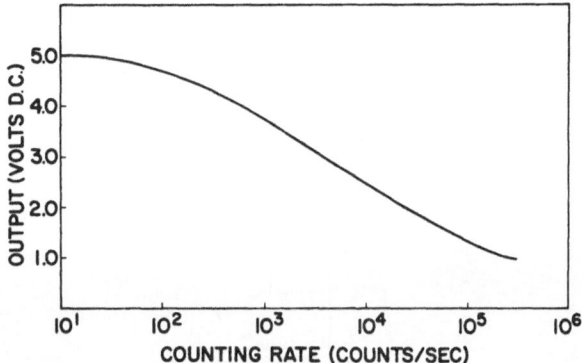

Fig. 5-4. Integrator calibration curve (reproduced with the permission of IEEE, from J. B. Reagan and R. V. Smith, IEEE, NS-10, January 1963).

cosmic rays, designed and developed a high-energy (> 50 MeV) gamma ray telescope [4], which was subsequently lofted on April 27, 1961, by the satellite Explorer XI. The design is a genuine example of applied physics and could serve as a stimulus toward a solution of design problems confronting the reader.

In the course of their work toward understanding the origin of the universe, Kraushaar and Clark focused their attention on gamma rays with energies greater than 50 MeV. These rays are important because they are usually created when high-energy particles collide with other particles. Being electromagnetic waves, they would, upon arriving in the solar system, give hints as to their point of origin, since they are not, like cosmic particles, deflected by magnetic fields they may have encountered on their way to the solar system. The objective of the Explorer experiment was to get far enough out of the earth's atmosphere that gamma rays produced by cosmic ray interactions with the atmosphere would not blot out the much fewer but very important galactic gamma rays. Further, by scanning the sky, a gamma ray map would be produced clearly showing the existence of high-intensity sources of high-energy gamma rays in the universe. This scanning was accomplished by an intentional tumbling of the satellite.

The detector shown in cross section in Fig. 5-5 utilizes the property of high-energy gamma rays to produce electron-positron pairs. For a gamma ray with energy greater than

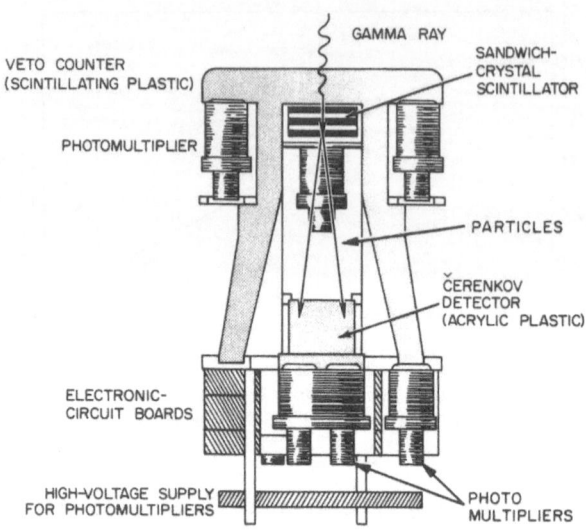

Fig. 5-5. High-energy gamma ray telescope (after W.L. Kraushaar and G.W. Clark, Scientific American, May 1962).

1.02 MeV the probability is high that the two particles, an electron and a positron, are created during a close encounter with a nucleus. The pair production occurs in a sandwich crystal scintillator made with alternating layers of NaI and CsI. The light pulse is then detected by the topmost center photomultiplier. The particles continue and eventually enter a block of Lucite plastic, which is part of a Čerenkov counter. A charged particle will generate light in a transparent substance if it travels faster than light travels in the substance, thereby automatically establishing a lower threshold for particle energies in the Čerenkov counter. The combination of the scintillator and the Čerenkov counter makes a coincidence circuit, which limits the registration to gamma rays entering only from within a 17° cone. The center counters are surrounded by scintillating plastic monitored by several photomultipliers. These provide anticoincidence signals to signal the occurrence of charged-particle entrances rather than high-energy gamma rays. There is a much higher probability that the gamma rays produce pairs in the NaI and CsI crystals than in the surrounding plastics, since the probability of pair production is proportional to the square of the material's atomic number (Cs = 55, Na = 11, I = 53, O = 8, C = 6, H = 1).

The described detector was flown successfully and produced a wealth of information. For instance, it ruled out the model of the steady-state universe. Subsequently in 1965 an advanced version of the gamma ray telescope was flown on the Orbiting Solar Observatory [5]. The much more extensive electronics in this package contain 8000 components and yet dissipate only 0.5 W.

Neutron Phoswich

Measurement of the neutron decay rate in the upper atmosphere can lead to information about its connection with the source of electrons in the Van Allen belts. To detect these neutrons the Lockheed Missiles and Space Company designed and developed the Neutron Phoswich [1]. The detector shown in cross section in Fig. 5-6 is a scintillation detector. It contains four wafers of Li^6I surrounded by a plastic scintillator. Thermal neutrons when interacting with the lithium produce alpha particles. The decay time of light pulses for the Li^6I crystals is on the order of 2.2 μ versus 3 to 5 nsec for the plastic scintillator. This difference allowed the use of a single photomultiplier tube for viewing the combined light output. By having a short (less than 1 μsec) time constant circuit monitoring the anode of the PM, an RCA 7151 C, and a 6 μsec time constant on the next to the last dynode, discrimination is

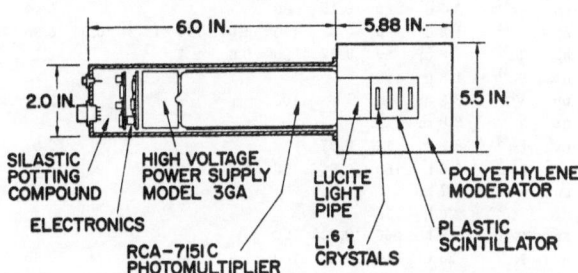

Fig. 5-6. Neutron Phoswich (reproduced with the permission of IEEE from J. B. Reagan and R. V. Smith, IEEE, NS-10, January 1963).

TABLE 5-II. Table of Activation Data for 14 MeV Miniature Neutron Generator

Element	Atomic wt.	Isotopic abundance	Reaction	Radioisotope	Half-life	Cross section (mb)	γ_1 Energy	γ_1 %	γ_2 Energy	γ_2 %	γ_3 Energy	γ_3 %	β_1 Energy	β_1 %	β_2 Energy	β_2 %
H	1	100	nγ		0		2.2	100	(Capture only)							
Li	6	7	np	He^6	0.8 s	6							3.5	100		
C	12	99	nnγ		0		4.4	90	(Inelastic scattering)							
N	14	100	n2n	N^{13}	10 m	5.7							*1.2	100		
O	16	100	np	N^{16}	7.4 s	49	6.1	55	7.1	20			10.4	24	0.4	54
Na	23	100	np	Ne^{23}	37 s	34	0.44	33	1.65	1			4.4	67	3.9	32
	23	100	nα	F^{20}	10.7 s	96	1.6	100					5.4	100		
Mg	24	79	np	Na^{24}	15 h	190	1 37	100	2.75	100			1.4	100		
	25	10	np	Na^{25}	60 s	45	0.58	10	0.97	15	1.6	6.5	4	65	3	25
	26	11	np	Na^{26}	1 s	50							5	100		
	26	11	nα	Ne^{23}	40 s	90	0.44	33	1.65	1			4.4	67	3.9	32
Al	27	100	np	Mg^{27}	9.5 m	79	0.83	70	1.0	30			1.75	70	1.6	30
	27	100	nα	Na^{24}	15 h	116	1.37	100	2.75	100			1.4	100		
Si	28	92	np	Al^{28}	2.3 m	220	1.78	100					2.8	100		
	29	5	np	Al^{29}	6.6 m	100	1.28	85	2.4	15			2.5	85	1.6	15
	30	3	nα	Mg^{27}	9.5 m	45	0.80	70	1.0	30			1 75	70	1.6	30
P	31	100	nα	Al^{28}	2.3 m	146	1.78	100					2.8	100		
	31	100	np	Si^{31}	2.6 h	86	1.3	0.07					1.5	100		
S	32	95	np	P^{32}	14 d	250							1.7	100		
	34	4	np	P^{34}	12 s	85	2.1	25					5.1	75	3.2	25
	34	4	nα	Si^{31}	2.6 h	125	1.3	0.07					1.5	100		
Cl	37	25	np	S^{37}	5 m	25	3.1	90					1.6	90	4.3	10
	37	25	nα	P^{34}	12 s	45	2.1	25					5.1	75	3.2	25
K	39	93	n2n	K^{38}	7.5 m	10	2.2	100					*2.7	100		
	41	7	np	A^{41}	110 m	80	1.3	99					1.2	99	2.5	0.9
	41	7	nα	Cl^{38}	37 m	30	2.2	16					4.8	53	1.1	31
Ca	42	6	np	K^{44}	12 h											
	44	2	np	K^{44}	22 m	75										
Ti	46	8	n2n	Ti^{45}	3 h								*1.0	100		
	47	8	np	Sc^{47}	3.4 d	50	0.16	60					0.44	60	0.6	40
	48	73	np	Sc^{48}	1.8 d	90	0.99	100	1.04	100	1.32	10.0	0.65	100		
V	51	100	np	Ti^{51}	6 m	27	0.32	96	0.6	4			2.1	95	1.5	5
	51	100	nα	Sc^{48}	1.8 d	29										
Cr	52	83	np	V^{52}	3.8 m	78	1.4	100					2.5	100		
Mn	55	100	nα	V^{52}	3.8 m	52.5										
Fe	54	6	n2n	Fe^{53}	8.9 m	10	0.37						*2.5	100		
	56	92	np	Mn^{56}	2.6 h	150	0.84	100	1.8	30	2.1	20	2 8	50	1.0	30
	57	2	np	Mn^{57}	1.7 m								1	100		
Co	59	100	nα	Mn^{56}	2.6 h	35										
Ni	58	68	np	Co^{58}	9 h	560	0.025	100								
	58	68	n2n	Ni^{57}	36 h	40	1.36	100	1.9	14	0.13	14	0.85	86	0.7	11
	61	1	np	Co^{61}	1.7 h	180	0.07	100					1.2	100		
	62	4	np	Co^{62}	14 m		1.2	100					2.9	75		
Sr	88	83	np	Rb^{88}	18 m	18	1.85		0.9		1.4		5.3	78	3.6	13
Ba	137	11	nn'	Ba^{137}	2.6 m	>600	0.66	100								

TABLE 5-II (cont.). Reprinted with the Permission of IEEE from R. Monaghan et al., IEEE, NS-10, January 1963

Relative γ detection efficiency; 3 × 3 NaI	Relative % detectability (counts/sec) 750 mg in sample 3 × 3 NaI	Relative γ activation (counts/sec) 750 mg elemental samples; 7×10^6 neutrons/cm^2-sec											
		Irradiate 12 sec			Irradiate 2 min			Irradiate 20 min			Irradiate 200 min		
		Delay			Delay			Delay			Delay		
		5 sec	30 sec	2 min	30 min	2 min	10 min	2 min	10 min	30 min	10 min	30 min	120 min
10	1150	490	50	–	63	–	–	–	–	–	–	–	–
10	600	110	70	14	310	60	–	65	–	–	–	–	–
25	3500	1600	300	–	500	–	–	–	–	–	–	–	–
37.5	8000	1.3	1.3	1.3	13	13	13	132	130	130	1200	1150	1025
10	60	7	5.5	2.0	30	11	–	15	–	–	–	–	–
10	130	25	15	3	65	15	–	18	–	–	–	–	–
50	5000	65	63	55	650	580	330	3500	2000	450	2500	560	–
37.5	5000	–	–	–	0.9	0.9	0.9	90	90	90	715	700	650
25	5700	340	300	190	2400	1600	140	3300	300	–	–	–	–
25	140	3.7	3.5	3	30	25	10	100	45	6	50	7	–
25	40	0.5	0.5	0.4	5	4	2.5	24	14	3	18	4	–
25	3800	220	190	120	1300	900	80	2000	180	–	–	–	–
6	20	8	2	–	4	–	–	–	–	–	–	–	–
20	120	3.3	3	2.5	25	20	7	75	25	2	30	2	–
6	60	25	6	–	12	–	–	–	–	–	–	–	–
25	200	4	3.7	3.0	33	30	15	145	70	10	85	10	–
25	115	–	–	–	1.5	1.5	1.5	14	13.5	12	80	70	40
30	50	–	–	–	2	2	1.5	17	14	10	40	30	5
25	30	–	–	–	2	2	1.5	9	7	4	20	10	1
37.5													
25	70	–	–	–	–	–	–	–	–	–	2	2	2
75	3500	–	–	–	2	2	2	18	18	18	175	170	160
25	400	13	12	9	90	70	30	320	125	–	140	–	–
75	450	–	–	–	–	–	–	0.7	0.7	0.7	75	72	70
25	1000	40	35	25	300	220	50	650	150	–	150	–	–
25	800	30	27	20	240	180	40	500	100	–	140	–	–
37.5	15	–	–	–	2	1.8	1	10	5	1	6	1	–
30	2400	2	2	2	21	21	20	200	190	170	1340	1230	830
30	600	0.5	0.5	0.5	5	5	5	50	45	40	320	300	200
25	5400	1.4	1.4	1.4	14	14	14	140	140	135	1320	1300	1150
25	400	–	–	–	–	–	–	3	3	3	26	255	25
37.5	40	–	–	–	–	–	–	6	5	4	28	24	12
25													
7.5	40	–	–	–	3	3	2	20	15	7	30	13	0.5
25	400	19	17	11	150	100	12	220	27	–	29	–	–

accomplished between neutrons and charged particles. A discrimination of 94% against electrons was realized. The power consumption of the instrument is 0.9 W at 28 V DC.

Semi-Rad

The Semi-Rad is gamma ray rate detector designed and developed by S. Kronenberger of the U. S. Army Electronics Laboratories in Fort Mammoth, New Jersey. It consists of an evacuated chamber with a center electrode positively charged for the collection of secondary electrons generated as the gamma rays interact with the detector walls. It requires gamma rates in excess of about 10^6 rads (C)/sec to produce enough output current for amplification in the currently available commercial version of the device. It is manufactured by the Eon Corporation and Reuter-Stokes, and measures approximately 1 in. in diameter and 6 in. long. The output is linear with gamma dose rate and has a dynamic range of 6 to 7 decades. It is extremely useful in applications where fast response is a requirement, as the inherent response of the detector is less than 10 nsec, as opposed to ion chambers, which are limited by the gas reaction time of from 50 to 100 nsec or more.

Material Analysis by Neutron Activation

The nuclear instrument described here was designed and developed by Dresser Research, Tulsa, Oklahoma, for application aboard a roving moon vehicle [6]. It works on the principle of neutron-activating the sample material. By subsequently measuring the emitted gamma radiation, the material can be identified. The main components of the instrument are (1) a miniature 14 MeV neutron generator powered by a 120 kV electrostatic generator in combination with an ion acceleration tube, which produces neutrons at the rate of 10^8 neutrons/sec, and (2) a scintillation spectrometer. Table 5-II shows the characteristics of various elements after activation by the 14 MeV neutrons. From the table a signature can be found which would fit known elements.

The equipment was designed for space application, although the method of neutron activation analysis is also widely used for many purposes in industry and science.

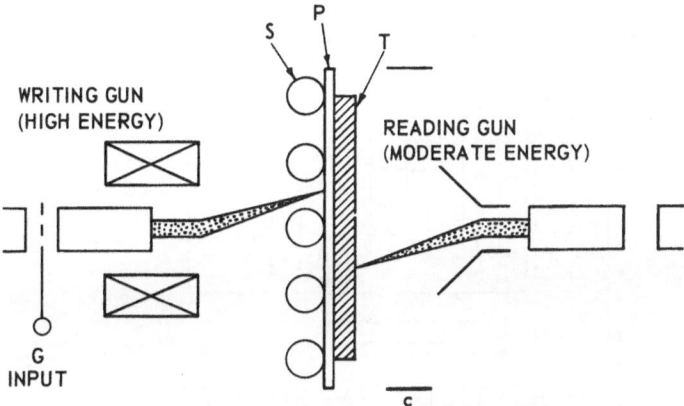

Fig. 5-7. Scan conversion storage tube (reproduced with the permission of IEEE, from S. W. Thomas, IEEE, NS-12, February 1965).

Airborne Scanner

Space telemetry confined to specific frequencies by IRIG standards often is found to provide insufficient bandwidth for transmission of space data. Such is the case when it is desired to transmit transient data such as that associated with a solar flare. To fill this need a design by Lawrence Radiation Laboratory, Livermore, California may find an application [7]. This design is based on the principle of scan conversion, and is capable of recording a single transient with a rise time of less than 1 nsec and reading out the data over telemetry with only 100 kHz bandwidth. The heart of the instrument is a scan conversion storage tube (Graphecon) shown in Fig. 5-7. The reading gun scans the target surface, knocking off secondary electrons in the process. These are immediately collected by C. The end result is a positive charge on T. The writing beam is of a high enough energy to pass through P and penetrate into T, where it causes induced conductivity in the dielectric. When the reading beam subsequently passes that point, secondaries are created for collection by C contributing to the signal output current. When this is displayed on a video screen, a picture will be scanned which is a replica of that introduced by the writing beam.

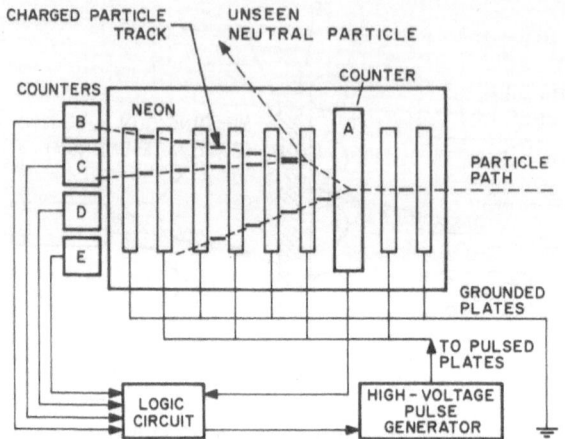

Fig. 5-8. Spark chamber (after G.K. O'Neill, Scientific American, August 1962).

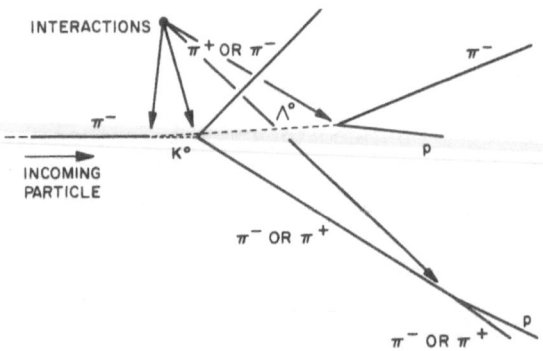

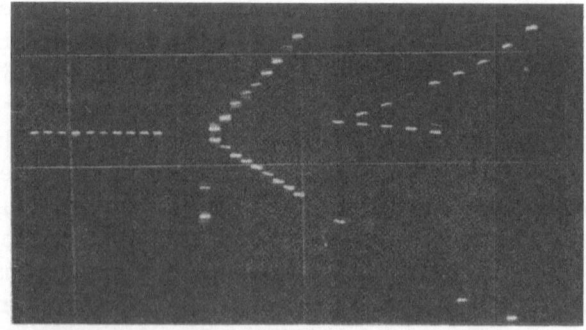

Fig. 5-9. Photograph of reaction (after G.K. O'Neill, Scientific American, August 1962; photograph was kindly provided by Dr. J.L. Cronin, Princeton University).

Nuclear Instruments for Particle Physics

The Spark Chamber. Of some 30 particles known today, as many as 16 were not known until in the late 1940's. This illustrates the increased rate of research in particle physics. In 1959 S. Fukui and S. Miyamoto of Osaka University, Japan, perfected the spark chamber, a particle detector, which proved to be an indispensable tool for the group of physicists from Columbia University and Brookhaven National Laboratory who in 1962 discovered the existence of two kinds of neutrino.

Devices used to detect elementary particles are divided into two broad classes, counters and track detectors. The counter simply registers and counts the passage of charged particles, whereas the track detector actually shows the particle track. The latter is important to the physicist who is interested in traces and time characteristics of all particles entering his detector. Then historically one moves from the cloud chamber in 1911, to the bubble chamber in 1950, and to the hodoscope chamber in 1955. Finally in 1957 two British physicists, T. E. Cranshaw and J. F. DeBeer, made the first strides toward a track detector which was perfected in 1959 by the Japanese physicists to become the spark chamber. It has an advantage over the bubble chamber (the other widely used track detector) in that it can be triggered.

The basic configuration of the spark chamber is shown in Fig. 5-8. It consists of an array of thin metal plates surrounded by a noble gas (neon or argon). A trigger device senses the arrival of the desired particle from the right and through the logic circuit applies a high voltage pulse to alternate plates in the chamber. Electrical sparks appear along the ionization trails left by charged particles, and in this manner outline their trails. In the figure, a charged particle interacts in the counter yielding one neutral and one charged secondary. The neutral secondary subsequently decays into one neutral and two charged particles. The counters at left keep count of created particles versus injected particles. Fig. 5-9 is a typical photograph of a reaction made by J. L. Cronin at the Brookhaven National Laboratory [8].

As the spark chamber is a rather new instrument, improvements are still being added. Immediately following the Japanese report, B. Cork of the University of California and J. L. Cronin of Princeton University had, respectively, 6- and 18-gap spark chambers in operation. G. K. O'Neill of Princeton University has concentrated on design of thin plate spark chambers which are useful for the study of charged particles in a magnetic field [8]. The Nuclear Science Group of the Institute of Electrical and Electronics Engineers continually conducts conferences where progress in the field of spark chambers is discussed [9].

   Semiconductor Particle Detectors. The quest for high resolution in their measurements has led low-energy particle physicists to the semiconductor particle detector as a necessary tool in their work. From the classical ion chamber, which can barely resolve to less than 10 μsec, the search for a better detector led first to the crystal counter. It used simply a block of semiconductor such as silicon across which a potential was applied. As radiation ionized the material atoms, a current commenced to flow in the outside circuit, which was a replica of the radiation pulse shape. Two drawbacks plagued the crystal counter—dark current and polarization.

   Dark current produced due to the finite resistivity of the silicon was the main problem in the early silicon detectors. As the techniques have improved for purifying silicon and accurately doping it to achieve a good balance between dark current and polarization, currently available crystal counters are of much better quality. Finally, the advent of the semiconductor diode or pn device was a tremendous improvement. The low dark current was in this device achieved by back bias and not by doping, which would enhance polarization. A back-biased diode has at the $p-n$ junction a so-called depletion region, which is free of carriers. As an ionizing particle traverses the junction area, the created free electrons are immediately (< 10 nsec) swept out into the external circuit, registering the signature of the radiation pulse. This is the effect that is termed the transient ionization effect in nuclear hardening of electronics. The time resolution therefore is a thousand times better than the ion chamber.

   Most semiconductor particle detectors used today are of the $n-i-p$ variety. The n and p layers are separated by a pure intrinsic layer of semiconductor material. Many of these are manufactured by the lithium drift method developed by E. M. Pell at the General Electric Research Laboratory. These are used for the particles, such as protons and deuterons, which have a low rate of energy loss per unit distance traveled.

SYSTEM DESIGN – MISSIONS IN SPACE

The nuclear instrument has a lasting place in space instrumentation systems. The research space flights flown during the past years rely on nuclear instruments to increase our knowledge about radiation in space. In the years ahead nuclear instruments will still be in demand for the continuing research of solar and galactic space. They will also be standard equipment aboard manned spacecrafts for the purpose of reporting to the crew on the state of the environment they are traversing.

Two typical nuclear instrumentation systems are described below. One is the instrumentation system aboard the Mariner II, the spacecraft which in December 1962 transmitted to the earth scientific data from within 22,000 miles of the planet Venus. The other is a description of the instrumentation system aboard the various spacecrafts and satellites designed to carry observatories outside the earth's atmosphere.

Mariner II Instrumentation System

The Mariner II spacecraft completed the first successful interplanetary voyage on December 14, 1962, when it passed within 22,000 miles of the planet Venus. It had traveled 180.2 million miles in 190 days, almost 1 million miles a day. The day of encounter with the planet, the radio signals from the spacecraft had to travel 36 million miles, introducing a delay between transmission and reception of 3.25 min [10].

Scientific instruments accounted for about 10% of the total weight of the craft. During the voyage scientific data were transmitted regarding magnetic fields, cosmic rays, and solar wind. Figure 5-10 shows two views of the spacecraft. The Mariner measured 5 ft in diameter at its base and 10 ft in height. Its wings containing solar cell panels measured 16.5 ft from tip to tip. Six cases mounted on the structure of the spacecraft contained the instrumentation, communication, and control systems. Of these, two contained the storage batteries

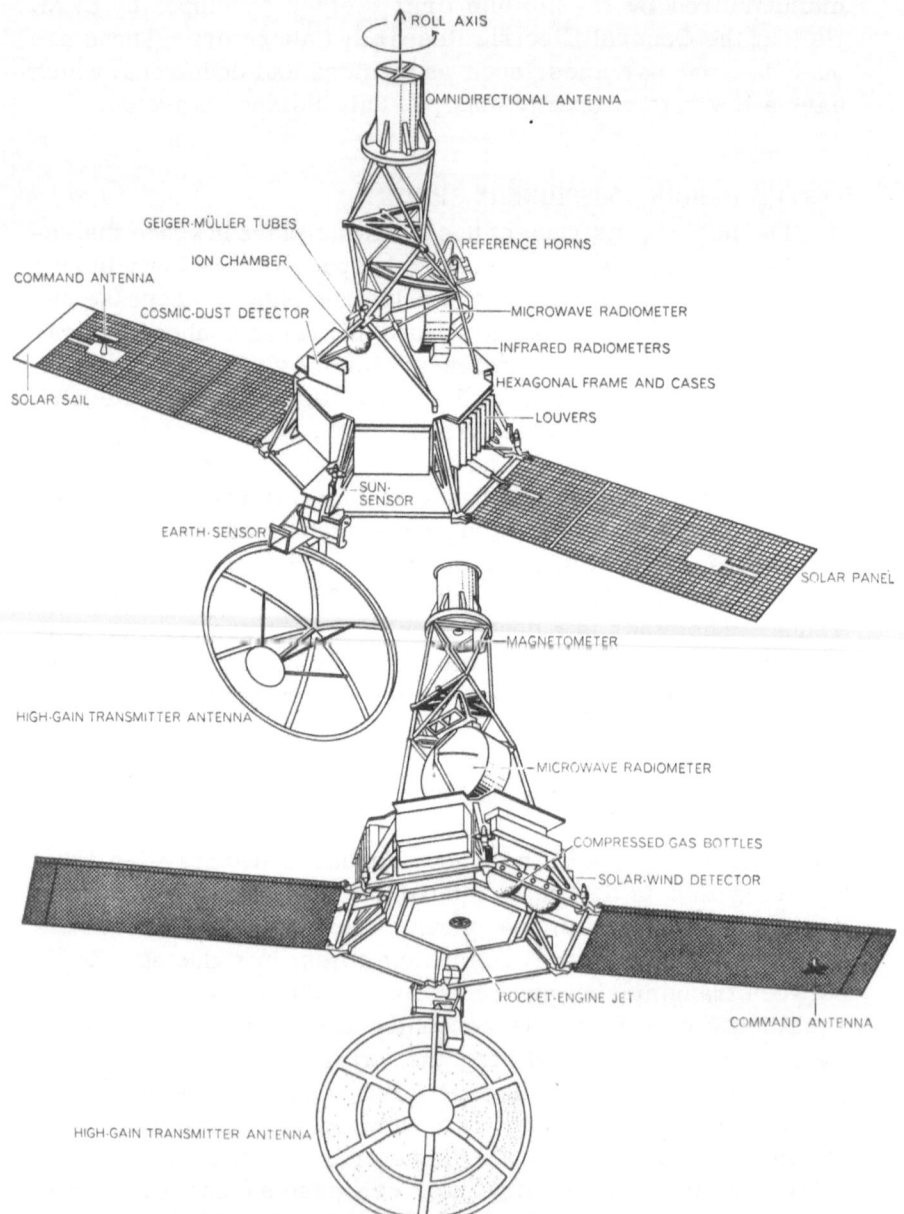

Fig. 5-10. The Mariner II spacecraft (reproduced with the permission of Scientific American, from J. N. James, Scientific American, July 1963).

and their regulator system, two held the communication system
(a receiver and a 3-W transmitter), one contained the signal
conditioning equipment, and one contained navigation equipment.
The cylindrical assembly on top of the spacecraft was an
omnidirectional antenna for transmission of its 3-W signal at
times when its directional antenna mounted in the opposite end
of the spacecraft was not pointing toward earth. To assure that
most transmission to the earth would be performed by the
directional antenna with its much higher gain, the spacecraft
was equipped with the earth sensor. The earth sensor mounted
at the hinge point of the directional antenna consisted of a
photomultiplier which tracked the reflected sunlight from the
earth. It was put into operation a week after launching, since
the earth was too bright for earlier operation of the earth
sensor. Subsequently, earth lock was lost once, possibly due
to a collision with some interplanetary matter, but recovered
within 3 min. The reception of signals from earth was accomplished by the small command antenna shown in Fig. 5-10
on one of the wings.

Directly below the omnidirectional antenna a magnetometer
was mounted. It consisted of three magnetic cores aligned along
the x, y, and z axes of three-dimensional space. The instrument was sensitive to 10^{-5} of the magnetic field on earth,
which ranges from a maximum of 50,000 gammas at the poles
to a minimum of 30,000 gammas at the equator. The magnetometer, designed by P. J. Coleman, Jr., and C. P. Sonett of NASA,
L. Davis of California Institute of Technology, and E. J. Smith
of JPL, reported that only weak magnetic fields exist in interplanetary space (64 gammas maximum) and a small magnetic
field exists near Venus (320 gammas maximum). In the middle
frame of the spacecraft can be seen a set of Geiger—Müller
tubes. Along with the ion chamber mounted next to them, these
instruments reported that the density and energy spectrum of
the high-energy cosmic rays from outside the solar system
remains constant regardless of the distance from the sun.
The experiment was designed by H: R. Anderson of JPL, H. V.
Neher of California Institute of Technology, and J. A. Van Allen
of the State University of Iowa. It measured protons greater
than 10 MeV, alpha particles greater than 40 MeV, and elec-

trons greater than 0.5 and 0.04 MeV. Total radiation dose collected for the trip was 3 roentgens, which is approximately what a person receives in 30 years on earth. To report on particles emitted by the sun was the objective of the solar wind detector. It was mounted on the lower part of the space-craft. The experiment was designed by M. M. Neugebauer and C. W. Snyder of JPL. It was designed to measure protons in the energy range from 240 eV to 8.4 keV. It reported on 40,000 spectra of particle energies throughout the flight and found that the solar plasma constantly pervades the inner interplanetary space. Designed as a spectrometer the instrument reported that the density of particles varies from 10 to 20 particles/in^3. On 20 different occasions the particle energies increased, indicating solar disturbances.

Mounted next to the ion chamber, a cosmic dust detector can be seen in Fig. 5-10. It consists of a crystal microphone mounted in the center of a sounding plate, which makes it sensitive enough to respond to particles as small as 1.3×10^{-9} g. The experiment, designed by W. M. Alexander of NASA, detected only two particles during the trip to Venus.

As the spacecraft reached its destination, the main experiments were activated. These were a microwave radiometer and an infrared radiometer, both mounted in the frame of the spacecraft. The microwave radiometer was designed by A. H. Barrett, MIT, D. E. Jones, JPL, J. Copeland, Army Ordnance Missile Command, and A. E. Lilley, Harvard College Observatory. Operating on two wavelengths, 13.5 and 19 mm, it was hoped it could answer some long-standing questions. The question as to whether the atmosphere of the planet contained water would be settled if the 13.5-mm signal were absorbed. The 19-mm wavelength could detect the surface temperature of the planet through its thick cloud cover. The 19-mm experiment determined that it is the surface of the planet that is hot and not the ionosphere, as previously contended by some researchers. The Mariner experiment indicated that the planet surface has a temperature of about 800°F. The 13.5-mm experiment reported very little water vapor in the atmosphere.

The infrared radiometer designed by L. D. Kaplan, JPL, G. Neugebauer, JPL, and C. Sagan, University of California, operated on two wavelengths, 8.4 and 10.4 μ. Its primary purpose was to detect breaks in the Venus cloud cover. It found none, but did report that the top of the Venus clouds remains at a temperature of −30°F both on the night and the day side of the planet. The theory of a hot surface is therefore substantiated.

Table 5-III summarizes the Mariner II experiments and their results.

Observatories in Space

The year 1946 marked the first time man had a look at the universe from above the protective layers of his planet's atmosphere, which protects man to the fullest extent from harmful radiation originating in space, but at the same time hampers his search for knowledge about the universe. Figure 5-11 shows how narrow a slit of the electromagnetic spectrum is available to researchers when scanning the skies from an earth-based observation point versus the window available from a balloon- or satellite-borne observation point [11]. In 1946, R. Tousey and his group at the U. S. Naval Research Laboratory mounted a spectrograph in a V-2 rocket and above the atmosphere made the first recordings of the sun's ultraviolet emissions. Balloon-borne observatories were launched to altitudes of more than 15 miles in the late 1950's, by M. Schwarzschild of Princeton University.

Then on March 7, 1962, the first true space observatory was launched under the direction of J. C. Lindsay of the Goddard Space Flight Center. This spacecraft, designated OSO (Orbiting Solar Observatory), was built by the Ball Brothers Company and transmitted almost 1000 hr of solar data to earth. It reported for example on rapid changes in intensity of soft rays from the sun over very short periods of time, changing by as much as a factor of four in as short a time as 1 sec. So far no explanation can be found to indicate the existence of such a generator on the sun. Figure 5-12 shows the OSO vehicle [11].

Table 5-IV shows other space observatories, which have either been flown or are planned flights in the years ahead [12]. The OGO (Orbiting Geophysical Observatory) contained a cosmic ray spectrometer developed by D. A. Bryant, G. H. Ludwig, and F. B. McDonald of the Goddard Space Flight Center.

TABLE 5-III

Mariner II Experiments (reprinted with the permission of Scientific American, from J. N. James, Scientific American, July, 1963)

Experiment	Measurement ranges	Findings	Experimenters
Microwave radiometer	Wavelengths of 13.5 and 19 mm	Venus surface temperature of about 800°F on both dark and light sides.	A. H. Barrett, Massachusetts Institute of Technology; D. E. Jones, JPL; J. Copeland, Army Ordnance Missile Command; A. E. Lilley, Harvard College Observatory
Infrared radiometer	Wavelengths of 8.4 and 10.4 μ	Top of Venus' clouds at −30°F. No breaks detected in clouds. Cold spot found, indicating possible high surface feature.	L. D. Kaplan, JPL and University of Nevada; G. Neugebauer, JPL; C. Sagan, University of California at Berkeley
Magnetometer	Up to 64 gammas in interplanetary space, up to 320 gammas near Venus	Venus has little or no magnetic field, is rotating slowly or not at all. Weak, fluctuating solar magnetic fields found in interplanetary space.	P. J. Coleman, Jr., National Aeronautics and Space Administration; L. Davis, California Institute of Technology; E. J. Smith, JPL; C. P. Sonett, NASA

Ion chamber and Geiger–Müller tubes	Protons above 10 MeV energy, electrons above 0.5 MeV, alpha particles above 40 MeV. Directional tube counted protons above 0.5 MeV, electrons above 0.04 MeV	Total radiation dosage for whole trip: 3 rads (air). Cosmic ray flux fairly constant throughout trip, not changing near Venus.	H. R. Anderson, JPL; H. V. Neher, C.I.T; J. A. Van Allen, State University of Iowa.
Crystal microphone	Dust particles as small as 0.0000000013 g	Detected only two particles throughout voyage, none near Venus.	W. M. Alexander, NASA Goddard Space Flight Center
Solar-plasma spectrometer	Protons from 240 to 8,400 eV	Solar wind "blows" constantly, varies in intensity and temperature with events on sun.	M. M. Neugebauer and C. W. Snyder, JPL
Radio (tracking)	Employed three-watt transmitter, high-gain antenna on Mariner at frequency of 960 MHz	More precise measurement of astronomical unit, mass of moon and Venus, and location of tracking station on earth.	T. W. Hamilton, J. F. Koukol, N. A. Renzetti, D. W. Trask, and J. D. Anderson, JPL

TABLE 5-IV

Orbiting Observatories—A Physical Comparison (reprinted with the permission of Space/Aeronautics, from I. Stambler, Space/Aeronautics, September 1964)

Class	Gross weight (lb)	Experiment weight (lb)	Basic shape	Overall dimensions (in.)	No. of experiments carried	Power (W)
OSO	458 (OSO-1), 542 (OSO-B2)	230 (OSO-1), 210 (OSO-B2)	Nonagonal lower section (wheel); half-moon upper section (sail).	Height, 37; wheel diameter, 44 (including gas jets, 92).	13 (OSO-1), 14 (OSO-B2).	16 (OSO-1), 18 (OSO-B2)
AOSO	1350	250	Cylinder.	Length, 125;* span (across solar paddles), 240.	4–6	400
OGO	1000–1500	150–220†	Rectangular box.	Length, 68; (booms extended, 54 ft); width, 33 (solar panels); height, 33 (unfolded, 20 ft).	20–50 at present, up to 100 with Atlas-Centaur.	500
OAO	3600	1000	Hexagon.	Length with sunshades open, 169; width across hex flats, 80; (with paddles extended, 254).	2 first flight; 1 each next two flights.	1000

*With occulting disks retracted.
†Up to 650 lb if Atlas-Centaur booster is used.

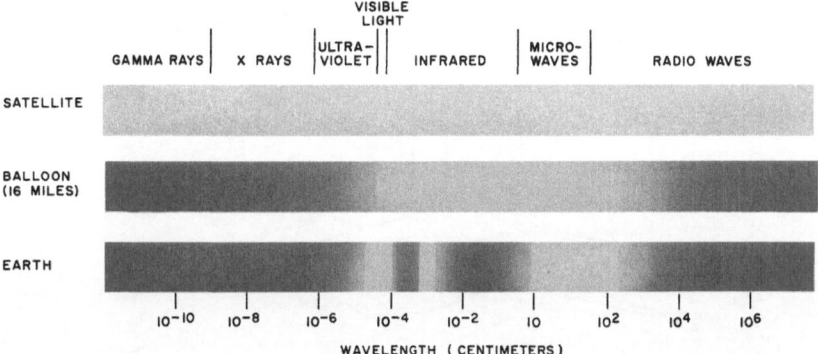

Fig. 5-11. Transmission of radiation through earth's atmosphere (after A.I. Berman, Scientific American, August 1963).

Three CsI (Tl) scintillators are used. The first is 0.15 g/cm² thick and 5 cm in diameter. The second CsI (Tl) crystal is 10 g/cm² thick and 5 cm in diameter. The third is an anticoincidence scintillator surrounding the second crystal on the sides and the bottom. The operation of the device is as described in the section on nuclear instrument design. The signals from

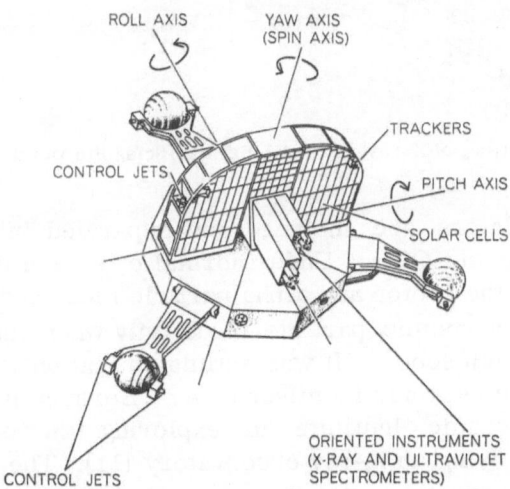

Fig. 5-12. OSO spacecraft (reproduced with the permission of Scientific American, from A.I. Berman, Scientific American, August 1963).

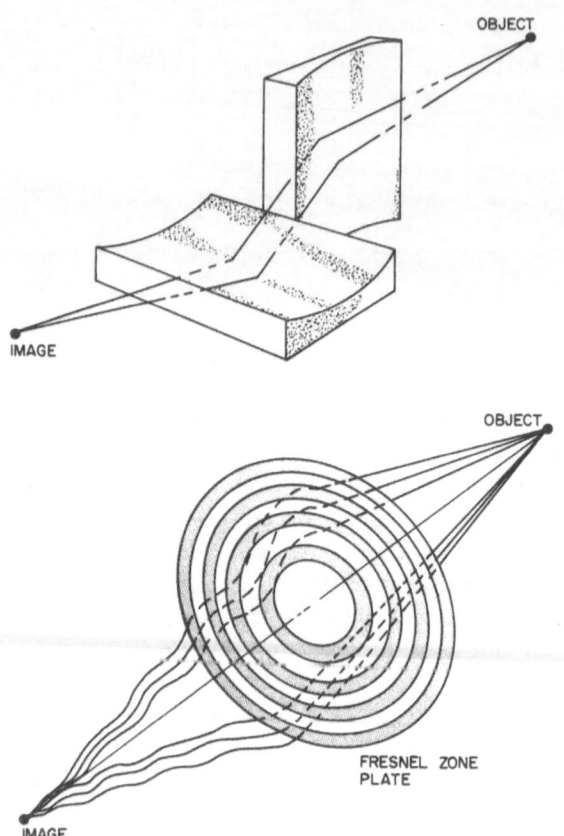

Fig. 5-13. Focusing of X-rays (after A.I. Berman, Scientific American, August 1963).

the scintillators are analyzed and separated into 256 channels aboard the OGO.   Under normal operation the analyzers
operate in the proton and alpha particle mode and switch over
to the heavy cosmic particle mode only when such a particle
enters the telescope.   It was estimated that only < 20% of the
heavy particles would be missed. A. I. Berman of the Rensselaer Polytechnic Institute is exploring the possibility of
eventually flying an X-ray observatory [11]. The advantage of
such an observatory over that of an optical observatory is the
increase in resolution made available by the decrease in observation wavelength. For instance, a point source of X-rays with

a wavelength of 5 Å will yield an image disk of only 0.001 diameter of the disk made by yellow light with a wavelength of 5000 Å. The difficulty in utilizing X-rays remains the problem of focusing them. Two methods are presently thought of as promising approaches, although they still require extensive development. Both were devised by A. V. Baez and are shown in Fig. 5-13.

In one method a narrow beam of X-rays from a point in space comes to a focus as result of having grazed off two cylindrical mirrors placed at right angles. This method was successfully tested using visible and ultraviolet radiation and it is therefore expected to work equally as well in the soft X-ray region of the spectrum. The second method, shown in the lower part of Fig. 5-13, is based on diffraction by means of a Fresnel zone plate, that is, a series of annular openings separated by opaque rings. At the center is a pinhole opening that establishes the area for each of the openings and rings. The X-rays pass through the annular openings but are absorbed by the rings made from gold foil. After passing through the openings, the photons of a given wavelength reinforce each other to create a disk image at a common focal point.

The orbit of an X-ray observatory should bring the spacecraft far enough into interplanetary space that radiation sources within the solar system are far enough removed for an acceptable signal-to-noise ratio to be maintained between these sources and the desired X-ray sources from outside the solar system. A. I. Berman recommends an orbit which takes the spacecraft as far as Jupiter or about 450 million miles from the sun. With the advent of nuclear powered spacecraft, orbits of this type should be readily attainable in the future.

REFERENCES

1. J. B. Reagan and R. V. Smith, "Instrumentation for Space Radiation Measurements," IEEE, NS-10, No. 1, 172–177 (January 1963).
2. J. H. Rowland et al., "Instrumentation for Space Radiation Measurements," IEEE, NS-10, No. 1, 178–182 (January 1963).
3. D. B. Hicks, L. Ried, Jr., and L. E. Peterson, "X-Ray Telescope for an Orbiting Solar Observatory," IEEE, NS-12, No. 1, 54–65 (February 1965).
4. W. L. Kraushaar and G. W. Clark, "Gamma Ray Astronomy," Scientific American 206(5):52–61 (May 1962).
5. F. Williams Sarles, Jr., and J. K. Roberge, "Low Power Fast Pulse Circuit Techniques, IEEE, NS-12, No. 1, 46–53 (January 1965).

6. R. Monaghan et al., "Instrumentation for Nuclear Analysis of the Lunar Surface," IEEE, NS-10, No. 1, 183-189 (January 1963).
7. S. W. Thomas, "Airborne Scan-Converter System," IEEE, NS-12, No. 1, 89-102 (February 1965).
8. G. K. O'Neill, "The Spark Chamber," Scientific American 207(2):36-43 (August 1962).
9. Ninth Scintillation and Semiconductor Counter Symposium, IEEE, NS-11 (June 1964).
10. J. N. James, "The Voyage of Mariner II," Scientific American 209(1):70-84 (July 1963).
11. A. I. Berman, "Observatories in Space," Scientific American 209(2):28-37 (August 1963).
12. I. Stambler, "The Orbiting Observatories," Space/Aeronautics 42(3):34-42 (September 1964).
13. L. Cramberg, "Fast-Neutron Spectroscopy," Scientific American (March 1964).

BIBLIOGRAPHY

Corliss, W. R., "Detecting Life in Space," International Science and Technology (January 1965).

Cranshaw, T. E., and J. F. DeBeer, "A Triggered Spark Chamber," Nuovo Cimento (May 1957).

Dearnaley, G., and D. C. Northrop, Semiconductor Counters for Nuclear Radiation, John Wiley & Sons, Inc. (New York), 1963.

Fukui, S., and S. Miyamoto, "A New Type of Particle Detector: The 'Discharge Chamber'," Nuovo Cimento (January 1959).

Rossi, B., "High Energy Cosmic Rays," Scientific American (November 1959).

Shklovsky, J. S., Cosmic Radio Waves, Harvard University Press (Cambridge, Mass.), 1960.

Snell, A. H., Nuclear Instruments and Their Uses, Vols. I and II, John Wiley & Sons, Inc. (New York), 1962 and 1965.

Literature published by the Nuclear Science Group of the Institute of Electrical and Electronic Engineers, New York.

CHAPTER 6

Electronic System Design Techniques

INTRODUCTION

The design of a radiation-hardened electronic system is in many ways no different from the design of electronic systems which must perform correctly while exposed to the more standard environments, such as temperature, vibration, and humidity. The early recognition of shock, vibration, temperature, humidity, and acceleration as environments which would influence electrical performance of the electronic systems has resulted in a thorough understanding of how to design against these environments. The reader familiar with electronic design in the more standard environments should endeavor to apply good design techniques from these environments to that of nuclear radiation, when the techniques seem applicable. This is seldom possible, since the radiation environment is in most ways quite unique, though there are instances where the approach can be used. For example, the design of electronics for operation in high temperatures (100 to 600°C) will also result in an electronic system which can withstand large integrated particle doses without degradation. The application of failure probability analysis, which is quite common for the more standard environments, has been generally neglected in work associated with the nuclear radiation environment. The importance of the failure probability analysis cannot be stressed enough. The single or few test sample determinations of component radiation sensitivity may be adequate early in the design cycle of a particular electronic system. However, toward the end of the design cycle, large test sample determination is a must in order to establish high confidence in the component selection for production units of the electronic system.

The radiation-hardening process is based on the application of good, well-proven electronic design techniques modified by measures accounting for the effects of nuclear radiation on the electronic system. The design procedure may be broken into the following definite steps, each of which will be discussed:

1. Nuclear environment analysis.
2. Possible application of shielding with an analysis of its influence on dose and spectrum of the environment.
3. Analysis of the entire electronic system.
4. Analysis of radiation-sensitive key subsystems and circuits.
5. Design of circuits in these subsystems.
6. Radiation testing to provide data for eventual redesign.
7. Large sample testing to prove soundness of design approach.

The chapter is presented in three general sections:

1. The Design Procedure. The logical step-by-step approach to radiation-hardened electronic system design is given, which the reader may find useful as a guide for his design problems.
2. Design Discussion. Key steps from the previous section are emphasized and discussed in greater detail. Certain data are presented along with the methods for their use.
3. Design Examples. A few examples are presented in order to show typical application of the procedure outlined above. Some of the literature available and applicable to the problem of radiation-hardened electronic system design is listed and described in the reference section.

THE DESIGN PROCEDURE

The Nuclear Radiation Environment

The design of a radiation-hardened electronic system begins with an analysis of the nuclear environment to be encountered. It must be determined, as precisely as possible, what kinds of radiation are likely to be encountered, whether gamma rays or particles, or a combination. Further the total dose, the dose rates, and the radiation spectrum must be determined accurately. Uncertainties in any of these determinations should be stated in order that a judgment can be made as to the limits of the environment quantities. Chapter 1 provides data on the makeup of various external environments. Modifications produced in these environments by materials interposed between the environment and the electronic system must be calculated. These modifications are not only differences in total integrated dose, but also changes in radiation spectrum.

Shielding

In addition to shielding provided by the space vechicle structure, the possibility of shielding individual electronic subsystems should also be considered. Chapter 3 provides data useful for this task. Under certain circumstances, shielding might be found to provide a reduction by just that extra amount in total integrated dose which it had been impossible for a particular solid-state device to withstand. Under other circumstances, such as for instance the shielding against neutron and gamma radiation from a nuclear reactor in flight, it may be too costly to rely entirely on shielding as the means for electronic system survival. If partial shielding has been decided upon, the new total integrated dose and the new radiation spectrum at the electronic system should be calculated.

The Electronic System

It is important and should be stressed that radiation hardening of an electronic system begins at the system level. By this is meant that the electronics engineer with the above environment analysis in hand functionally designs the electronic system so that obvious nuclear radiation failure mechanisms are avoided on a system level. It may also be possible with knowledge of the environment and its interactions to use this knowledge in developing special design techniques. As an illustration of the first point, the electronics engineer may decide to base the system design on vacuum tube techniques rather than solid-state techniques to gain the additional nuclear radiation tolerance of vacuum tubes. As an illustration of the second point, consider the application of back-biased diodes at strategic points in the system which by short-circuiting prevents system failure during a pulsed radiation environment.

Radiation hardening of existing electronic systems begins with a system analysis for the effects of radiation on the various subsystems and circuits. However, some of the creative design techniques that are useful in a new system may not be as readily applicable. The following technique is often used by electronics engineers to arrive at a radiation-hard design, and it is generally quite valid. The circuit or subsystem is designed to perform its electrical function without regard for the nuclear radiation environment. The next step consists of a radiation environment analysis, after which the circuit is modified accordingly to operate correctly in the environment. The technique is recommended as a first-time approach, but as familiarity with the nuclear radiation as an environment increases, it should be possible to design directly for tolerance to nuclear radiation.

The first step in the system analysis is the establishment of failure criteria. There is the first obvious failure criterion, which is inability to perform the system function. This criterion may then be expressed in a number of ways with a different failure mechanism producing each failure. With the sensitive subsystems identified the design procedure can proceed to the next step.

The Circuit Analysis

With the radiation-sensitive subsystems and circuits identified, the analysis of these proceeds to determine means for making the circuit or subsystem perform as required in the radiation environment.

The circuit failure criterion is first established by setting the minimum output requirements and minimum allowable changes to circuit transfer characteristics. The immediate interfaces with other nearby circuits must also be examined, since the response of the circuit to radiation, especially high

dose-rate ionizing radiation, is often dependent upon the characteristics of nearby circuits.

The circuit analysis is performed using standard techniques
well known to electronics engineers with appropriate parameters inserted to represent the influence of nuclear radiation.
The Ebers–Moll, the Linvill, and the Charge-Control mathematical models are those most often used for transient
analysis [5].

Circuit Design

With the results of the above analysis, the actual hardware
design of the circuit may proceed. The actual circuit design
makes use of a combination of techniques, among which are
(1) selection of nuclear-hard electronic parts, (2) special
arrangement of the parts to form a more radiation-hard
circuit, (3) special design techniques to overcome compromises
forced by parts selection, (4) selection of nuclear-hard
insulation and other construction materials, and (5) special
packaging techniques to eliminate air in contact with circuit
surfaces. The section below expands on this subject and
presents the reader with data for this task. The task may
broadly be divided into (1) design against permanent damage
and (2) design against transient ionization effects. The first
task is approached by selection of radiation-hard electronic
parts or by using design techniques which automatically compensate for the degradation in parts performance. An example
of the latter is the design of a solid-state amplifier with a very
high open loop gain. By utilizing negative feedback across this
amplifier, its input-to-output gain is only slightly affected by
parts degradation for a large range of particle dose.

The second task is accomplished by selection of electronic
parts and by using special design techniques, up to ionization
rates from 10 to 10^9 rads (C)/sec. The effects produced in
circuits can often be likened to noise, and it is found that those
design techniques that produce circuits tolerant to noise will
also work quite well against transient ionization effects.

Where the circuit will be affected by both permanent and
ionization effects the design order is generally to design
against the permanent effects first and the ionization effects
second. However, this should not be applied as a strict rule,

since under certain circumstances the environment itself by its timing may dictate the design order.

Simulation of the Environment

Accurate simulation of the environment is seldom if ever possible. A thorough analysis is therefore always called for to correlate as well as possible the chosen simulation environment with the actual radiation environment. Out of this analysis can be derived the best suitable simulation radiation facility and the dosimetry required to best correlate the spectrum of the simulation facility with the actual spectrum. Chapter 4 should be consulted prior to selection of a simulation facility.

Radiation testing of electronic parts will often precede circuit design and data from this parts testing are often used to augment the circuit analysis. A large amount of parts test data are available in the literature. Some are listed at the end of the chapter. However, because radiation testing of electronic components is fairly new it is difficult to correlate the available test data. Radiation testing results for one environment may not correlate with the test results from a different type of environment, so care should be used when comparing performance on the basis of test data from the literature. The reader may want to make his own electronic parts tests. In that case the following procedure is generally followed:

1. A survey of available parts test data is made to ascertain whether a similar part is reported on (see reference data at end of chapter).
2. If none can be found, a small sample test (2 to 6 devices) is conducted on a number of candidate electronic parts (5 to 10 types).
3. From the test in (2) two favorite candidate types are selected, and a large sample test is conducted on each of these. The sample size should be determined statistically based on the expected failure mechanism, but it generally ranges from 20 through 50 samples of the same type electronic part.

The test results from this test are arranged to produce failure probability data in order to select the part which will produce the most reliable circuit design. A technique for failure probability analysis is described below. It is emphasized that the third step is an essential tool for the electronics engineer to enable him to guarantee a reliable circuit or subsystem design.

Radiation testing subsequent to the circuit design has a twofold purpose: (1) It is a useful tool in discovering circuit responses that the analysis missed. The circuit analysis in most instances will accurately predict the response of the circuit to radiation. However, it should be kept in mind that a breakover point does exist where the circuit analysis may be far too complex and too costly by itself and needs to be augmented by circuit radiation testing. The reader should use his judgment for each individual case, and always be completely aware of any built-in limitation in the chosen circuit analysis technique. (2) A second type of circuit radiation testing is similar to that of the large sample parts testing. It is intended to provide the designer with confidence that the circuit design is reliable to the desired degree. This type of testing should be emphasized whenever the circuit has been designed with only a small design margin with respect to the environmental levels.

Chapter 4 should be consulted for experimental procedures. Based on these the reader may prepare a standard procedure suited for any particular testing program. An important part of the program is a well outlined and adhered to reporting method, which will later enable accurate correlation with other radiation test data.

DESIGN DISCUSSION

Electronic Parts Selection

The nuclear radiation tolerance of a specific circuit is dependent upon its "weakest link." A complete material and parts selection is the best first approach to radiation hardening of a particular circuit or subsystem. As described earlier the interaction of nuclear radiation with materials is a direct interaction with the atomic and nuclear structure of the material. By analyzing the individual circuit materials and selecting those materials with most inherent hardness to nuclear radiation, the first step toward a radiation-hard circuit is assured.

Jones [1], Keister [2], and REIC Report No. 36 [3] contain great amounts of useful data in this respect. The literature from the Radiation Effects Information Center of Battelle Memorial Institute in particular continually provides updated information regarding radiation test data from industry. Table 6-I summarizes the permanent effects of gamma and neutron radiation on various electronic parts. A guide for the selection of electronic parts now follows.

Active Devices

Semiconductors. The reader is referred to Chapter 2, which contains design equations for calculation of the permanent degradation of semiconductor devices and calculation of the induced currents by ionizing radiation. These calculations and small sample size radiation testing provide the tools for selecting the candidate devices on which to base the early circuit design. The final certification of the semiconductor devices should only be based on statistical testing providing failure probability data.

Vacuum Tubes. These may for the purposes of this discussion be considered as belonging to either of two types: (1) glass filamentary vacuum tubes or (2) ceramic filamentary vacuum tubes.

A third and special type is the heaterless ceramic vacuum tube. This vacuum tube and its complementing components (capacitors and resistors) belong to a family of electronic component parts developed by the Tube Department of the General Electric Company. Termed the TIMM (Thermionic Integrated Micro Module), the component family combines the nuclear hardness of ceramic vacuum tubes with the small power consumption of transistor circuits. One of the design examples below describes a design that was based on the exploitation of these advantages.

Glass Filamentary Vacuum Tube. The glass filamentary vacuum tube can be considered in two groups: (1) vacuum envelope of borosilicate glass and (2) lead glass envelope tubes.

Mechanical damage in the first type takes the form of fracturing the envelope as stresses are produced during boron capture of thermal neutrons. Glass-to-metal seals also fail. As shown in Table 6-I, these failures limit the usefulness of this type of tube to particle doses equivalent to 10^{15} to 10^{16} neutrons/cm^2. Then if glass tubes are used it is advisable to avoid boron-containing glass envelopes.

In the glass vacuum tube family, one finds an interesting device which has been created specifically for application in transient ionization radiation hardening work. It is called the CIRCUITRON. It is made by Sylvania and is an entire electronic vacuum tube circuit (flip-flops, amplifier, etc.) mounted inside a 12AU7 type non-boron glass envelope. The placement of all circuit elements of a particular circuit in a vacuum minimizes the ionization effects by about 50% as compared to ordinary vacuum tube circuits. The nuclear hardness of the device and other non-boron glass vacuum tube circuits falls at about 2 to 5 $\times$ 10^{16} neutrons/cm^2 (>10 keV). Outgassing from tube parts is primarily responsible for this threshold.

Ceramic Filamentary Vacuum Tubes. Ceramic filamentary vacuum tubes have been tested to 10^{17} neutrons (>10 keV)/cm^2 without failure, and it is reasonably estimated that their upper limit is between 10^{18} to 10^{20} neutrons (>10 keV)/cm^2.

It is obvious that by using vacuum tubes, the electronic system becomes very nuclear-hard. On the other hand, the power requirements for vacuum tube designs eliminate these from consideration in most space vehicle system applications. It was this limitation that prompted the creation of the heaterless ceramic vacuum tube. Its virtue is the fact that as the amount of active devices in an electronic system grows larger than about 500 to 1000, the system power requirement decreases below that of a comparable filamentary vacuum tube system, and for quite large systems (greater than 1000 active devices) the power requirements begin to approach those of a transistor system.

Transient ionization effects in vacuum tubes also consist of ionization-generated currents. The major contributor is secondary electron emission from envelope and electrodes. The current only persists during the ionization pulse. The

TABLE 6-I

Radiation Sensitivity of Various Electronic Materials (reproduced with the permission of Electronics, from H. L. Olesen, Electronics, January 1965)*

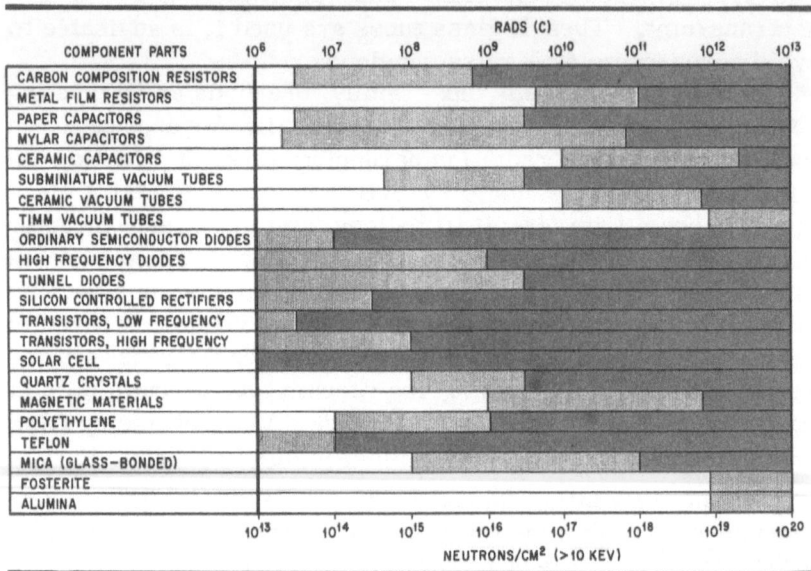

*White area, unaffected; hatched area, slightly affected; solid area, severely affected.

effects of air ionization are more severe in vacuum tube circuits than in transistor circuits because of the higher circuit impedances required in the vacuum tube circuits.

Transient ionization effects in vacuum tubes can be calculated from the following equation [6]:

$$i_{g R} = V_0 / G R_g \dot{\gamma} \qquad (6\text{-}1)$$

where $i_{g R}$ is the change in grid current, V_0 is the output transient voltage, R_g is the grid resistance, $\dot{\gamma}$ is rads/sec, and G is the amplification factor. The basic mechanism causing the ionization effects is ejection of secondary electrons from the tube parts, envelope, and electrodes. As the equation implies, the observed current in the circuit is a function of grid resistance. The choice of a tube should be made based on the following rules: (1) Select one with materials of a low secondary electron emission efficiency, such as titanium. (2) Use as low impedances as feasible in the external circuits; the grid resistor especially should be of low value. (3) Remove air from exposed terminals by either encapsulation of the terminals or evacuation of the circuit container.

Studies have shown that certain ceramic materials have smaller ionization responses at high temperatures. This fact is a benefit for the cases where a TIMM system can be used.

Passive Devices

Resistors. Resistors, when compared to the active devices, are quite resistant to permanent degradation. As Table 6-I indicates, it is generally better to select metal film resistors for the cases where high integrated doses are expected. For extremely high doses, where passive devices are required to complement ceramic vacuum tubes, the passive components from the TIMM family may be of use.

The ionization effects on resistors may be shown by a shunt leakage resistance and an injection current generator, and expressed mathematically by

$$R_s = A/\dot{\gamma} \qquad\qquad (6-2)$$

$$i_c = B/\dot{\gamma} \qquad\qquad (6-3)$$

where A, given in Ω-roentgen/sec, is $\geq 10^{12}$ for an unencapsulated carbon composition resistor and $\geq 4 \times 10^{13}$ for an encapsulated carbon composition resistor and metal film resistor. B is given in A-sec/roentgen and ranges from 10^{-12} to 10^{-15} dependent on encapsulation and radiation spectrum. Keister [2] should be consulted for more detailed data or an experiment should be arranged for the particular resistor and the constructional features of the circuit. The resistor types above are surpassed from a permanent damage standpoint by the wire-wound resistors. The hardness of this type of resistor is limited only by the bobbin material. By selecting inorganic material such as ceramic, these resistors will provide the passive complements for the ceramic vacuum tubes. For environments where surface effects are expected to occur, it is also advised to use wire-wound resistors, since the surface

construction of the metal or carbon film resistors is likely
to be affected by the surface effects.

Capacitors. The selection of capacitors for least permanent
damage can be focused on the selection of the capacitor
dielectric. If the dielectric is inorganic, such as glass or
ceramic, the capacitor is quite resistant to permanent effects.
The inorganic dielectrics are affected at low doses when
compared to active device performance in Table 6-I, and may
not be usable in many instances.

The ionization effect response of capacitors can be shown
as a radiation-induced shunt leakage resistance, R_s

$$R_s = \frac{1}{\sigma_{ns} C} \qquad (6\text{-}4)$$

where σ_{ns} is the induced conductivity consisting of a prompt
and a delayed term, as follows:

$$\sigma_{ns} = \sigma_p + \sum_i \sigma_{di} \qquad (6\text{-}5)$$

where the prompt term is

$$\sigma_p = K_p \dot{\gamma} \Delta \qquad (6\text{-}6)$$

where K_p is a constant from 10^{-5} to 10^{-3} determined by the
dielectric type, $\dot{\gamma}$ is gamma dose rate, and Δ is a constant
0.5 to 1.

The delayed conductivity is produced by the presence of
long-lived traps in the dielectric, i.e., traps which hold the
radiation-generated carriers for times exceeding the disap-
pearance of the ionization pulse. The traps have different time
constants and are each assigned a separate value for the
following term:

$$\sigma_d = K_d \dot{\gamma} \Delta t_p \exp(-t/\tau) \qquad t_p < \tau \qquad (6\text{-}7)$$

where t_p is the ionization pulse width and τ is the trap lifetime.
Capacitors are serious producers of ionization effects in electronic circuits, more so than any of the active devices. They
form in combination with resistors RC time constants, which
prolong the effects beyond the termination of the ionization
pulse. Generally try to avoid capacitors in circuits where
ionization effects are undesirable, or else limit their capacitive

values, since as shown in equation (6-4) the shunt resistance R_s is a function of capacitance. Keister [2] also points to a possible choice of a favorable dielectric with respect to reduction of ionization effects.

Magnetic Materials. Generally magnetic materials will outperform semiconductor devices, and are therefore useful design components in solid-state systems. Beyond the solid-state boundary for radiation survival one must distinguish between soft and hard magnetic materials. The hard magnetic materials such as Alnico have thresholds for damage as high as 10^{20} neutrons (> 0.1 keV)/cm^2. Table 6-II shows the effects of radiation and temperature on various magnetic materials.

Failure Analysis

A great deal of the radiation test data that are available for the task of electronic parts selection tend to indicate a 50% failure probability, because experimenters were looking for gross effects on parts in terms of an average value of radiation exposure. These data are helpful to the electronic system designer early in the design cycle of a particular system, when the broad parameters of the system are arrived at. However, as the design proceeds this type of data becomes very unsatisfactory, since the electronic system designer is required to specify his system in terms of a 99.99% success probability. At this point in the design cycle entire electronic circuits could be tested to produce this type of reliability data, and sometimes are. Parts data given in terms of failure probability are also used to analytically determine the probability of success for the circuits containing the parts.

F. W. Poblenz [4] has shown how the Weibull distribution can be applied successfully to the task of predicting the probability of failure for parts irradiated by nuclear radiation. The relationship between p, percent survival, and the wearout life is given by

$$\frac{p}{100} = \exp\left[1-\left(\frac{x}{\theta}\right)^b\right] \qquad (6\text{-}8)$$

where p is survival, x is the wearout life (neutrons/cm^2), θ is the scale parameter, and b is the slope parameter. The

TABLE 6-II

Effects of Radiation and Temperature on Various Magnetic Materials

Magnetic material	Radiation stability (10^{16} fast neutrons/cm^2)	Temperature stability (500°C)	Probable stability in combined environment
Magnetic cores			
Silicon—iron (unoriented)	VG	VG	VG
Silicon—iron (grain oriented)	VG	G	G
27 Cobalt—iron	—*	G	G
35 Cobalt—iron	—*	G	G
2 V Permendur	VG*	G	G
Supermendur	—*	P	P
16 Aluminum—iron	VG	NU	NU
2 Molybdenum—thermenol	—	NU	NU
12 Aluminum—iron	—	P	P
Supermalloy	NU	NU	NU
4-79 Molybdenum—permalloy	NU	NU	NU
Mumetal	NU	—	—
50 Nickel—iron	P	NU	NU
50 Nickel—iron (oriented)	P	NU	NU
Ferrites	VG	NU	NU
Powder cores	VG	NU	NU
Permanent magnets			
Alnico 5	VG*	VG	VG
Alnico 6	—	VG	VG
Alnico 2	VG*	G	G
Alnico 3	—	G	G
Alnico 1, 4, and 7	—*	U	U
Remalloy	—	U	U
Indalloy	—	U	U
Vicalloy	—	U	U
0.9 Carbon steel	—	U	U
Alnico 12	VG*	NU	NU
Tungsten steels	—	NU	NU
Chrome steels	VG	NU	NU
Cobalt steels	VG*	NU	NU
Cunife 1	VG	NU	NU
Cunico	VG*	NU	NU
Barium ferrite	VG	NU	NU

Notation: VG, very good; U, usable; P, poor; NU, not usable.
*Becomes radioactive.

fraction F failed at x neutrons/cm^2 is defined by

$$F = 1 - \frac{p}{100}$$ (6-8)

Then combining with equation (6-7),

$$F = 1 - \exp\left[-\left(\frac{x}{\theta}\right)^b\right]$$ (6-10)

The Weibull distribution then becomes a straight line plot on Weibull probability paper, as shown in Fig. 6-1. The slope of the line is given by b, which is therefore called the slope parameter. It signifies the failure rates of the parts being irradiated, as follows:

$b = 1$: Failure rates are constant and the Weibull distribution is exponential.

$b < 1$: Failure rates are decreasing.

$b > 1$: Failure rates are increasing.

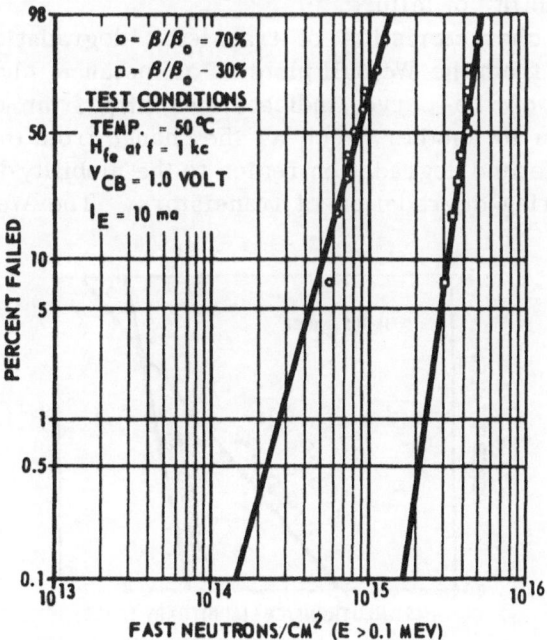

Fig. 6-1. 2N797 transistor, percent failed versus neutron dose (reproduced with the permission of IEEE, from F. W. Poblenz, IEEE, NS-10, January 1963).

It will be noticed from Fig. 6-1 that the curves have data points only in the high percentage failure region, but may then accurately be extended to the low percentage failure region. Note also that the slope for the $\beta/\beta_0 = 30\%$ curve is much steeper than the $\beta/\beta_0 = 70\%$ curve. This is a clear indication that the particular transistor is approaching the threshold region of its degradation curve. Figure 6-1 shows a transistor which is still in the stable regions (minority carrier lifetime or mobility regions) of its degradation curve.

The importance to the electronic system designer of this type of reliability test data is emphasized by Fig. 6-2. In this figure two Weibull distribution curves are shown, one for transistor A and one for transistor B. Note that if a selection in this case had been made based on a 50% probability of failure, transistor A would clearly have been chosen. This would have caused grim consequences in terms of system reliability, which must be based on 0.1% probability of failure figures. As Fig. 6-2 shows, transistor B is far superior to transistor A at 0.1% probability of failure.

Other characteristics of transistor degradation can be discerned from the Weibull plots. For instance, changes in b, the slope of the curve, indicate switching from one failure mechanism to another, such as the change from the minority carrier lifetime degradation region to the mobility degradation region during degradation of transistors. The Weibull plots

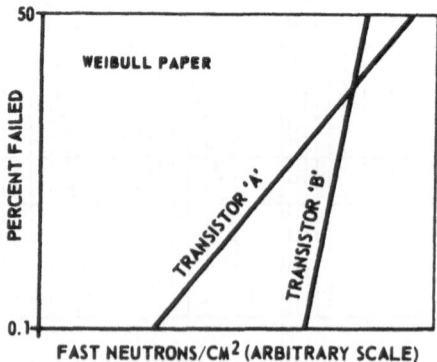

Fig. 6-2. An effect of the statistical nature of failure (reproduced with the permission of IEEE, from F. W. Poblenz, IEEE, NS-10, January 1963).

also indicate the quality of the device manufacturing process. The more uniform the process the steeper the slope of b. This in terms of degradation means the device performs well in radiation until drastic changes in the crystal structure take place and that the degradation is not controlled by surface phenomena.

The application of the Weibull distribution is of course not limited to transistors, but is a powerful tool for describing the probability of failure for all electronic parts. The method is highly recommended, and it is the author's hope that it may become standard practice to express radiation test data in probability terms for the benefit of data users.

Transient Analysis

Prediction of electronic circuit response to transient ionizing radiation has become common practice among electronic system designers. The electronic circuits are expressed in terms of their equivalent circuits, and appropriate shunt resistances and current generators are inserted at the appropriate locations in the equivalent circuits to represent the ionizing radiation input. The equivalent circuit models are checked against the actual response of the hardware circuits in linear accelerators and flash X-rays. Subsequent to this, certain values in the equivalent circuit model may be changed in order to account for actual responses in the test. The equivalent circuit models are subsequently used to predict response to transient ionizing radiation over a wide range of ionization rates.

The analysis method early in the design cycle generally consists of hand calculations using standard circuit analysis techniques such as Ebers–Moll, Linvill, etc., modified to include radiation parameters. Later in the design cycle computers are used in order to increase the accuracy of the analysis. A number of computer codes are available for use by electronic system designers. The reader is advised to become familiar with the limitations of a particular computer code before using it. For instance, any code that automatically truncates its intermediate results could present the user with an erroneous impression of actual conditions if the user was

unaware of the nature of the code. Most of the codes available
are digital, but for readers who have analog computers, transient analysis can also be performed on these. The analog
computer, however, is best for situations that do not demand
wide dynamic ranges of variables.

A number of examples on transient analysis are found in
Sullivan and Wirth [5]. Information regarding the availability
of computer codes may be obtained by those qualified from the
Radiation Effects Information Center, Battelle Memorial Institute, Columbus, Ohio.

It should be emphasized that the use of computers and other
analytical tools is limited to application on existing circuit
designs. Early in the design cycle it is broad knowledge and
creative skill on the part of the electronic system designer
that determine the system design. The computer and other
analytical tools are subsequently used to refine and confirm
the design.

Packaging Design

Packaging of electronic circuits needs careful attention in
situations where transient ionizing radiation is expected. In
order to confine the produced ionization it is common practice
to exclude all air from circuit surfaces by either of two
methods: (1) encapsulation of solid-state circuits by a well-
adhering compound such as silicone or (2) evacuation to less
than 1 mm Hg for circuits where the thermal output does not
allow encapsulation.

DESIGN EXAMPLES

The following examples are short, but do represent the
type of reasoning that lies behind radiation hardening of
electronic circuits. Three examples are given: digital
circuit design, analog circuit design, and TIMM record amplifier. Two are solid-state designs, whereas the third is an
example of a design for particle doses in excess of those that
a solid-state circuit can withstand.

Digital Circuit Design

Neutron Damage.   Before a circuit can be designed for reliable operation the effect of nuclear radiation on end-of-life characteristics is studied.   Once the worst-case effects on every component are known, a worst-case circuit design analysis may be performed.   Figure 6-3 shows a typical NAND gate that is used here as an illustration of worst-case design technique.   The resistors are carbon composition, and all diodes are silicon, which can tolerate approximately 10^{16} neutrons/cm^2.   Diode leakage current will not seriously affect operation of the circuit unless it exceeds 120 μA. The most serious problem is the effect of low current gain on the circuit fan-out.   The fan-out for a design center transistor is 20. After taking into account worst-case effects other than nuclear radiation, the worst-case fan-out is approximately 5.   By further reducing the fan-out, the circuit can be re-designed to tolerate a total integrated neutron flux of 10^{13} neutrons/cm^2. To increase the fan-out or the radiation tolerance, two solutions are possible: Replace the transistor with a device that is more radiation resistant, or increase the power dissipation of the circuit.   Replacing the single transistor with a Darlington configuration will increase the fan-out and radiation resistance, since overall current gain is approximately the product of the individual transistor current gains. Figure 6-4 shows the improved circuit schematic.

Although the worst-case design will ensure reliable operation, it is often an inefficient technique.   The number of

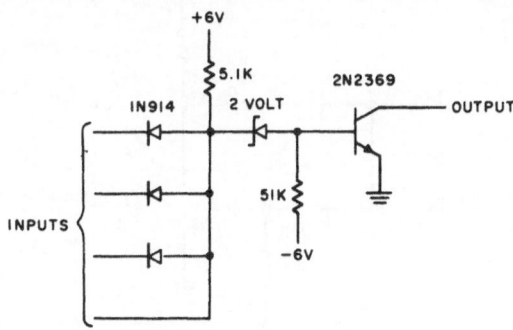

Fig. 6-3. Typical NAND gate.

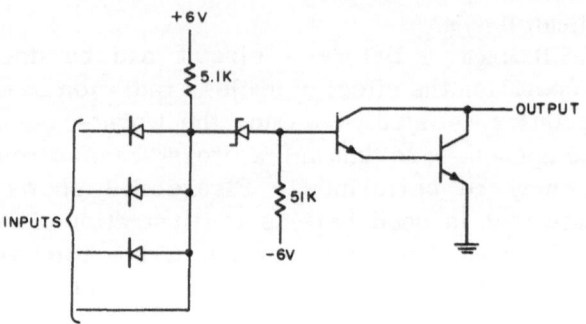

Fig. 6-4. NAND gate with improved neutron-dose capability.

components and the power dissipation are high, and the logical fan-out is low. Another technique is the use of probability theory, especially for the β parameter of transistors. This method shows the statistical effect of nuclear radiation on the components. By selecting the $3\,\sigma$ point, a circuit can be designed with greater fan-out and higher component tolerances. Investigations into the statistical effect of nuclear radiation on the transistors should be undertaken. The sample size of transistors and the other parts for determination of statistics generally runs from 20 through 100 units. Probability calculations described above can be employed after these investigations are complete. By using a feedback path that minimizes the effect of transistor β on a circuit, an increase in the radiation tolerance may be obtained. For example, the circuit shown

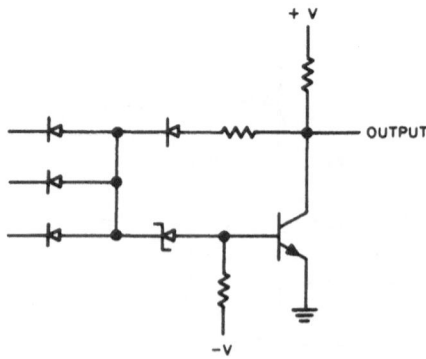

Fig. 6-5. Unsaturated gate using feedback to stabilize operation.

in Fig. 6-5 has a feedback loop; the output voltage is not affected by any reasonable change in the value of β and the fan-out of the circuit is still a function of the β of the transistor.

Pulsed Gamma Tolerances. Experiments at the Sandia Pulsed Reactor Facility have shown that conventional logic circuits are capable of tolerating 10^7 rads (C)/sec with the pulse width approximately 50 μsec. Many devices are assumed able to tolerate the higher levels of pulsed nuclear radiation, but the final answer depends upon the effect on the overall system. Presently digital semiconductor circuits can be built with a pulse effect threshold of 10^9 rads (C)/sec.

Transient radiation effects from a gamma pulse can be quite similar to a noise pulse in an electronic circuit. If a circuit is tolerant of noise pulses, the probability of malfunction due to transient radiation-induced pulses will be low. One method for reducing the effects is to minimize the impedance of the various elements; the difficulty in lower impedance circuits is the increased power dissipation. Another technique is to increase the switching threshold voltage.

If it is feasible to reduce the operating speed of digital circuits so that the propagation delay is greater than 1 μsec, it is possible to minimize the transient gamma radiation problem. This is based on the assumption that the transistor is the most vulnerable component and that the effect on the other components is not significant. Gamma pulses often cause the transistor collector current to continue flowing for as long as 1 μsec after the transient. The pulse of collector current may appear as a voltage transient which is propagated from

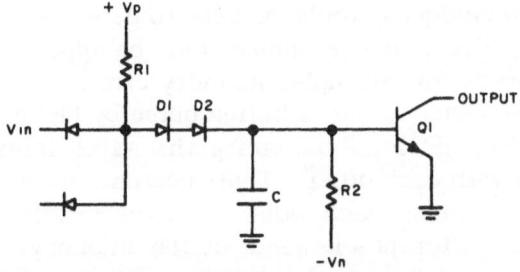

Fig. 6-6. Digital NAND gate with capacitive noise filtering.

circuit to circuit unless a bypass filter capacitor is added. The maximum radiation that the circuit will tolerate thus becomes a function of the filter network. An example of a digital NAND gate with capacitive filtering is shown in Fig. 6-6. Referring to the figure and assuming that the driving (preceding) transistor is saturated, a gamma pulse would tend to cause it to go deeper into saturation. The input voltage V_{in} would be nearly zero, if the driving transistor were initially off, but saturated after the gamma pulse. V_{in} would then change from approximately V_p to nearly zero. Diodes D_1 and D_2 would be back-biased until capacitor C discharged; the capacitor supplies base current to the transistor Q_1 while the input voltage is low. The time constant c times the transistor input resistance must be greater than the time duration of the input voltage transient under worst-case conditions. The value of c thus would be approximately 1 μF, which is not compatible with integrated circuit techniques.

A digital system could be designed to tolerate a high level of pulsed gamma radiation by using memory or storage techniques. The logic circuits in such a system could assume any arbitrary state during the transient, but the system would use the information stored in hardened registers or memories and return afterwards to normal operating conditions. A detector would be necessary to indicate when the logic circuits are not operating properly. If a sonic delay line were utilized as the storage medium, a technique known as "tapping the line" could be considered (i.e., the information being processed by active devices would also be stored in a sonic delay line). In this redundant method, a failure of the active device would not cause the information in the delay to be lost. Majority logic and voting techniques would be used to determine the correct information. The same technique could be applied to magnetic memory cores; for example, memory circuits could be made almost impervious to the radiation pulse by having two circulating memory devices containing the same information but out of phase with each other. Their contents would be checked periodically against each other. When a pulse arrived, it would affect different elements of the memory. During the next comparison interval, these differences would be noted and automatically corrected.

Digital System Considerations. Although it is possible to design individual circuits to tolerate high levels of neutron and gamma radiation, other considerations must be given to the overall system. It has been shown that power dissipation increases as the nuclear radiation total dose and dose rate are increased, but the relationship is not linear.

Adequate protection against radiation implies increased size and weight in many systems, especially for pulsed radiation where capacitor filtering and redundant techniques may be necessary. The power consumption of a system using ceramic tubes would be considerably higher than one using semiconductor circuits. Adding to the disadvantages these devices require a period of time for warm-up and stabilization.

Analog Circuit Design

General Considerations. Amplifier circuits operating in a radiation environment are subject to two specific phenomena which may cause malfunction. One is transients, which certainly will cause erroneous information; the other is permanent damage which may change the circuit transfer characteristics. Both of these phenomena must be considered in the design of hardened circuits.

Investigation of failure modes for various types of components justifies the use of the following design criteria: (1) metal film resistors, (2) ceramic capacitors, (3) high-frequency transistors for all applications, and (4) diodes made so there is no void at the junction, and (5) operation at low voltage levels and high current levels with low impedance base-bias current.

Voltage Amplifiers. In order to preserve the transfer characteristics of a circuit, the design should be such that the parameters of the active devices are restricted to second-order terms (insignificant terms) in the transfer characteristic equations. Consider the amplifier circuit illustrated in Fig. 6-7 as an example.

This circuit can be used as a DC amplifier or as an AC amplifier by coupling in and out with a capacitor. The circuit has excellent bias stability because of the DC feedback and it can be shown that the gain is essentially independent of the transistor parameters. The voltage gain of a common emitter amplifier with un-bypassed emitter resistor is approximately

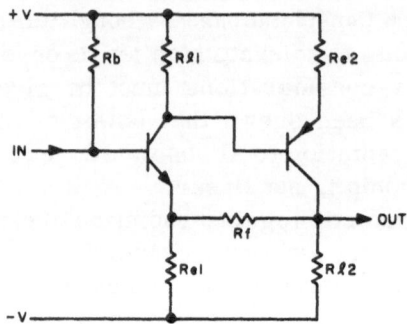

Fig. 6-7. Schematic of single-ended, analog voltage amplifier.

equal to the collector load impedance divided by the emitter impedance (for moderate stage gain). Therefore, if we minimize interstage loading, the forward gain of the amplifier is approximately

$$A = A_1 + A_2 = \frac{R_{L1} \; R_{L2}}{R_{e1} \; R_{e2}}$$

The loop gain with feedback is

$$A' = \frac{A}{1 + \beta A}$$

where

$$\beta = \frac{R_{e1}}{R_{e1} + R_f}$$

Therefore the gain with feedback is

$$A' = \frac{R_{L1} R_{L2} / R_{e1} R_{e2}}{1 + (R_{e1} / R_{e1} + R_f)(R_{L1} R_{L2} / R_{e1} R_{e2})}$$

$$= \frac{R_{L1} R_{L2}}{R_{e1} R_{e2}} \left(\frac{R_{e1} + R_f}{R_{e1} + R_f + R_{L1} R_{L2}} \right)$$

which is independent of the transistor parameters. This is a desirable feature because the mode of failure of a transistor under radiation is a decrease in h_{FE}, the forward gain. Another desirable feature of this amplifier is that leakage current will subtract because of the use of complementary transistors. It should be noted that under pulsed radiation there is an increase in leakage currents. This curciut will not suppress transients; however, by using high-frequency transistors the duration of these transients can be minimized because they will be a function of the base lifetime of the carriers. Transients in amplifier circuits, if short in duration, should not present major difficulties since no memory function is performed by the amplifier.

Better transient suppression and good small-signal charac-teristics can be obtained by using a difference amplifier operated in a single-ended fashion. The circuit illustrated in Fig. 6-8 which was tested in the Godiva reactor is slightly unconventional in that low impedances are used and the tran-sistors are operated at relatively high currents.

Q_3 is a constant current generator whose current is determined by R_1, R_2, and R_3. The operating current is selected to be considerably higher than the peak currents ex-

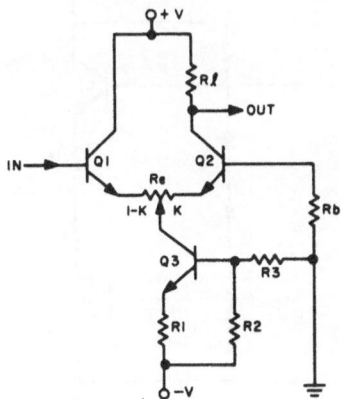

Fig. 6-8. Schematic of balanced type analog voltage amplifier.

pected from radiation transients. In this case the common mode rejection features of the circuit will cancel out transients, provided Q_1 and Q_2 have matched characteristics and do not saturate. The base of Q_1 must be terminated in an impedance equal to R_b. The gain of this circuit is approximately

$$G = \frac{R_L}{kR_e + R_b/\beta}$$

The gain of the circuit can be made to be almost independent of the transistor parameters; i.e., $R_e \gg R_b$.

This circuit showed sufficient transient suppression in the Godiva test to indicate acceptable operation at gamma rates of 6.75×10^6 rads (C)/sec. Operation at higher levels probably could be obtained. The trade-off would be circuit gain. Drift and offset voltages, always a problem in differential amplifiers, would have to be compromised. However, low values of R_b and matched transistors with respect to V_{be} tracking would minimize these problems. The two amplifier circuits being discussed are low impedance circuits. It is suggested that emitter followers be used for impedance matching. A comple-

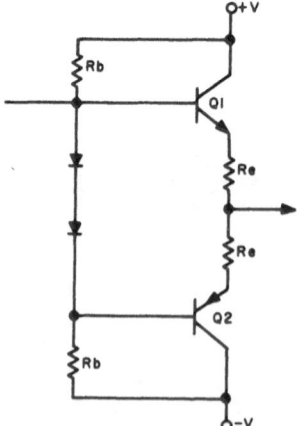

Fig. 6-9. Schematic of a complementary emitter follower.

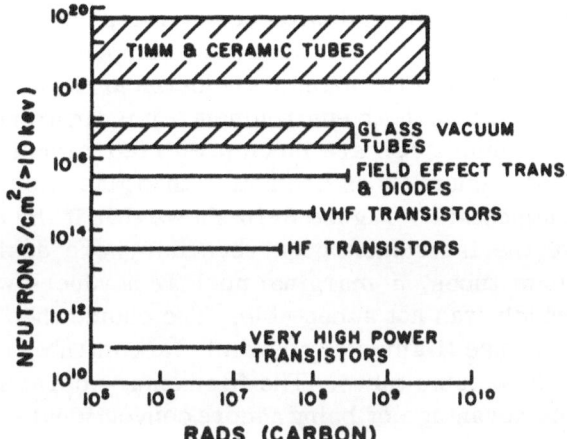

Fig. 6-10. Nuclear hardness.

mentary type emitter follower should be used to take advantage
of its common mode rejection capabilities, as shown in
Fig. 6-9. Leakage currents and transients cancel this circuit.
However, the impedance transfer characteristics are functions
of the transistor parameters, particularly the gain B of the
transistor.

TIMM Record Amplifier

The nuclear-hardness specifications for the record amplifier were the following: (1) no degradation of system performance for a neutron dose of less than 10^{16} neutrons (>10 keV)
/cm^2 and a gamma ray dose of 10^6 rads (C), and (2) rapid
recovery from a short-pulse transient ionizing radiation at
10^{12} rads (C)/sec rates.

For most applications, where a nuclear hardness is
required and specified, packaging materials and passive components can be obtained which exceed the nuclear hardness of
active components by several orders of magnitude. To achieve
the specified nuclear hardness one must then choose an active
component which will maintain its performance even after

being subjected to the radiation doses specified. Figure 6-10 compares a number of available active components with regard to their capability to maintain electrical and mechanical performance in a nuclear environment. It is apparent from the graph that vacuum tubes are much preferred to transistors and that ceramic vacuum tubes are better than glass vacuum tubes.

TIMM Components. Figure 6-10 shows that if the electronic portions of the instrumentation recorder were designed with glass vacuum tubes, a marginal nuclear hardness would have resulted which was not acceptable. The choice then remained as whether to use filamentary ceramic vacuum tubes or heaterless ceramic vacuum tubes. The filamentary ceramic vacuum tube had the advantage of being a more conventional component. The heaterless ceramic vacuum tube was on the other hand specifically developed by the General Electric Company, Receiving Tube Department, Owensboro, Kentucky, to provide a family of vacuum tubes, which (1) were more nuclear-hard, (2) consumed less power, and (3) were of less weight and volume than the conventional filamentary ceramic vacuum tubes. The nuclear hardness was achieved by constructing all components (triodes, diodes, resistors, and capacitors) from ceramic and metals. The power consumption for electronic systems using 500 or more active components was reduced as shown in Fig. 6-11 by a factor of about 10. This was accomplished by excluding the heater from the individual vacuum tube and instead supplying the heater power required from an external source, which would heat the entire electronic system to 580°C. Figure 6-11 indicates the power consumption of an operating TIMM system. Approximately three times this power is required to start a cold TIMM system. The size and weight of the individual TIMM component is only one tenth that of a conventional ceramic tube, and Fig. 6-12 compares the weight of a filamentary vacuum tube system with that of a TIMM system. The large initial weight for the filamentary tube system is a result of the basic power supplies which are required for this system to provide filament power.

The term TIMM is the acronym for Thermionic Integrated Micro Module circuits. The heart of the system is the heaterless ceramic vacuum tube in which cathode emission is

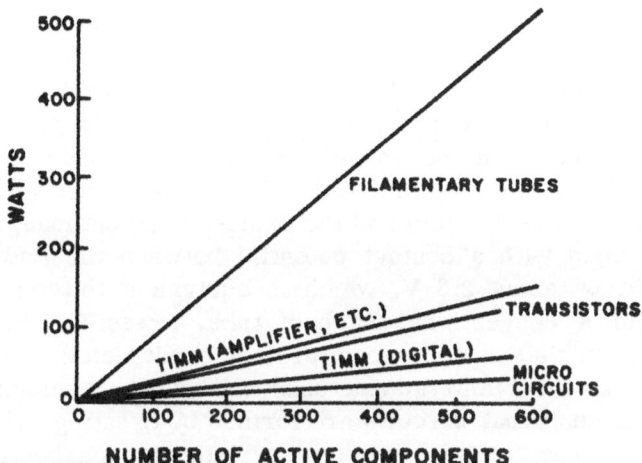

Fig. 6-11. Power consumption.

obtained by heating the entire device to a temperature of
580°C. This device, together with diodes, resistors, and small
capacitors, constitutes the component family from which
TIMM systems are built.   The TIMM diode is a device
consisting of three parts, a fosterite ceramic ring and two
titanium plates. The cathode is mounted on one of these plates.
It consists of a thin wafer of platinum foil which is oxide-

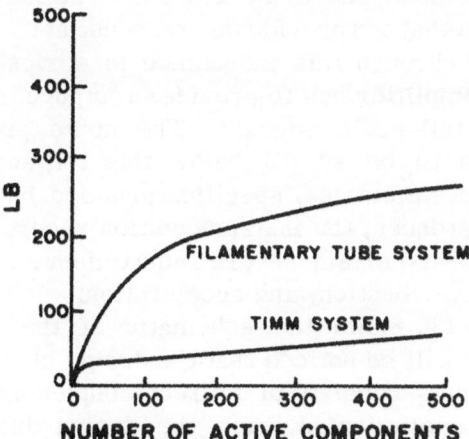

Fig. 6-12. System weight.

coated.   The anode is simply the clean titanium face of the opposite plate. The TIMM triode also has a planar configuration of elements.   The cathode is a platinum disk coated with barium—strontium oxide emissive material.   The grid, spaced a few thousandths of an inch from the cathode, is a mask etched from thin titanium foil.   Due to the difference in work function of the titanium and the oxide-coated cathode, the tube is provided with a contact potential between the grid and the cathode of about 2.3 V, which is equivalent to the zero bias point of a conventional vacuum tube.   Passive components such as resistors and capacitors are constructed in the same manner as the TIMM vacuum tubes, and have the same diameter ($\frac{5}{16}$ in.) such that circuits are formed by creating cylindrical stacks.   The TIMM resistor consists of a carbon filament deposited on a ceramic substrate which is mounted in the evacuated space between two titanium electrodes.   The capacitors are stacked ceramic metal plate structures. Figure 6-15 showing the TIMM record amplifier indicates the construction of the various components.

Record Amplifier.   The amplifier is required to amplify millivolt signals from instrumentation sensors such that these signals can be recorded on the magnetic tape. The signals will contain frequency components as high as 200 to 250 kc and the upper 3 dB point on the response curve for the instrumentation recorder therefore had to be 250 kc.   The inductance of the magnetic head that can provide this response at 60 ips tapespeed is 0.1 mH.   Through this inductance in series with a 1000 Ω resistor the amplifier has to provide an output current of 2-mA rms for a full-scale signal.   The noise produced by the amplifier has to be 40 dB below this full scale signal.   In addition to these electrical specifications and the requirements for nuclear hardness, the instrumentation recorder and amplifier had to meet certain severe requirements regarding mechanical shock, vibration, and acceleration.

Figure 6-13 shows the schematic of the TIMM record amplifier.   It will be noticed that the design chosen to achieve the requirements consists of a direct-coupled amplifier. Capacitors were eliminated from the design due to the large physical size of the coupling capacitors required to achieve the

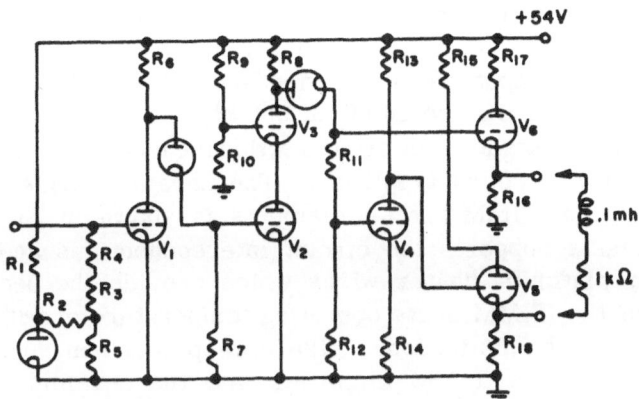

Fig. 6-13. TIMM amplifier schematic.

low end of the band pass and due to the longer recovery time from transient radiation effects inherent in circuits with large RC time constants. The obvious coupling problems which are incurred when trying to couple from the high voltage point of the plate to the grid of the succeeding tube were solved by using a diode-resistor divider network. The coupling diodes have two functions: (1) to provide optimum bias for V_2 and V_4 when used in conjunction with resistors R_7, R_{11}, and R_{12}, and (2) to provide AC coupling between these stages without impeding low frequencies. The direct coupling feature permitted the pass band of the amplifier to be flat down to DC. To achieve the 250 kc end of the band pass was more difficult. The Miller effect of the TIMM tubes and an inherent capacitance of the TIMM resistors due to their small size cause high-frequency roll-off. The Miller effect was minimized by using low value resistors in grid return circuits, as in V_2, V_4, V_5, and V_6, by reducing the stage gain, as in V_1, or by grounding the grid, as in V_3. The output stage is a balanced push-pull stage to meet the requirement that no DC current could be permitted to flow in the magnetic recorder head and thereby cause an erroneous bias signal. To further reduce the effects of transient radiation which develop best across very high impedances the highest value resistor used in the design is 63 kΩ. The electrical performance of the amplifier is linear for input voltages from 0 to 15 mV. It provides a current gain of 40 dB. The noise produced by the amplifier is 42 dB below a F. S. output. Its input impedance varies with frequency being 100 kΩ at DC and 20 kΩ at 250 kc.

Figure 6-14 shows in cross section the mechanical features of the TIMM record amplifier. The ceramic chassis which supports the TIMM circuit elements is shown in Fig. 6-15. The chassis supports the circuit interconnections and holds the two platinum heater wires which provide the heating to maintain the TIMM at its operating temperature of 580 ± 3°C. The chassis is 2.63 in. long, 1.125 in. deep, and 1 in. high. The assembly of the TIMM amplifier and the ceramic chassis constitute the TIMM module. Figure 6-14 shows how the TIMM module is located in the center of the cylindrical container and supported by thermal insulating material. This material is a quartz wool packed to a density of 8 lb/ft^3. The little heat that is transmitted to the outer edge of this material is reflected back into the center by a 7-layer thermal shield made from dimpled molybdenum foil. The cylindrical container is constructed from stainless steel and a heater is permanently welded to one end of the cylinder. The entire assembly is evacuated to a pressure of 10^{-7} mm Hg. The major function of the vacuum is to eliminate the transient radiation effects which occur as gamma radiation ionizes air in contact with circuit surfaces and thereby increases the leakage currents

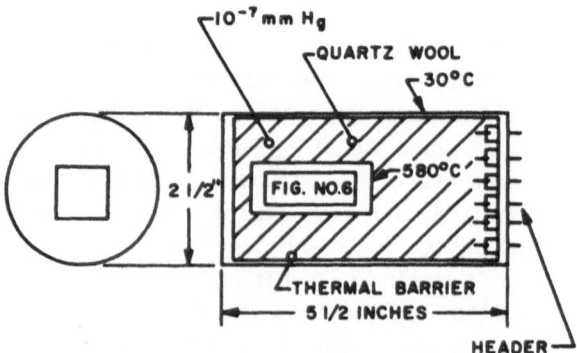

Fig. 6-14. Mechanical features.

across these surfaces.   Organic construction materials were avoided, since these have inherently long recovery times from transient radiation.

The heating power required for this package once the thermal lag of the materials has been overcome is 5 W. To overcome the thermal lag requires the application of 57 W for 10 min or 9 W for 2 hr. The heater was designed for a voltage of 7 V.

Radiation Test. A radiation test of the TIMM amplifier was performed by the Air Force Weapons Laboratory in the SPRF reactor, Albuquerque, New Mexico. This radiation test reactor has been used for many similar experiments by companies throughout the country, such that the results from the TIMM amplifier test can be judged against the results from all these other experiments.   The TIMM components can be thought of as belonging to the family of vacuum tubes. Whenever other vacuum tube circuits have been tested in the SPRF reactor, the circuits became inoperative for intervals of from 100 μsec to several milliseconds after the gamma radiation pulse.   The testing of the TIMM amplifier marks the first time a tube-type circuit has operated during the SPRF gamma radiation pulse with such slight effects.

The radiation levels during the test were: gamma dose rate, 3×10^7 rads (H_2O)/sec; gamma dose, 10^3 rads (H_2O); neutrons, 8×10^{12} neutrons (>10 keV)/cm^2.

From an instrumentation trailer a 100 kc sine-wave signal was fed to the TIMM amplifier inside the reactor facility. The 100 kc signal was amplified by the TIMM amplifier, and fed back out to the instrumentation trailer where it was recorded on a 500 kc tape recorder. During this test only a slight base-line perturbation was visible. The 100 kc sine-wave remained intact throughout the test. As the frequency of the test signal was increased to 200 kc, the portion of the sine-wave lying within the base-line shift disappeared for a period of 100 μsec. This indicates a bandwidth sensitivity to pulsed gamma radiation.

The gamma pulse from the SPRF reactor has a 50 μsec width, and should therefore fall well within the electrical bandwidth of the TIMM amplifier.   From the standpoint of

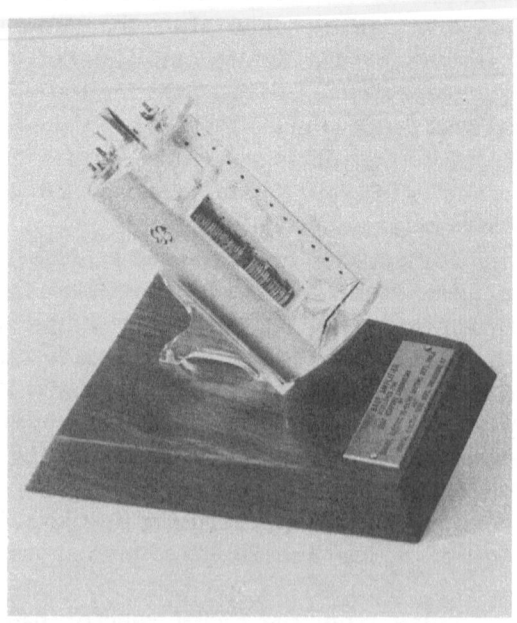

Fig. 6-15. TIMM amplifier package.

gamma pulse width this test will probably be the most severe
the TIMM amplifier will be subjected to.

Pneumatic Controls and Logic
Pneumatic circuits as well as controls are feasible. They
do not have any moving parts and use fluids to control fluids.
Many electronic functions can be performed analogously with
these devices. For example, they can simulate multivibrator,
flip-flop, memory, logic analog, and amplifier functions. They
have been built into complicated servo-type mechanisms and
have successfully operated test vechicles. Their full potential
has not been developed. Several companies are known to be
investigating them independently. It appears to be an approach
to radiation-resistant circuits which should not be overlooked,
although some materials problems can also be expected.

Timers
Electronic timers employing RC time constants or semiconductors are quite vulnerable to radiation. Count-down
timers which store the integrated count in magnetic cores should
be less vulnerable to pulsed radiation than those which employ
semiconductor flip-flops, but it is difficult to eliminate active
devices and their vulnerability altogether.
Electric watch movements offer a possibility as a radiation-
resistant low-frequency device of moderate accuracy. One
of them, developed by Hamilton Watch Company, utilizes the
conventional balance wheel movement, except that the main-
spring is replaced by a battery, magnetic coils, magnets, and
a switch. This device might be suitable for long-term aerospace use, particularly if the battery could be solar-charged.

REFERENCES

1. D.C. Jones (ed.), TREE (Transient Radiation Effects on Electronics) Handbook,
Battelle Memorial Institute (February 1964); obtain from DDC, Doc. #AD432213.
The TREE Handbook is an excellent source of design information for hardening
electronic systems against transient radiation. It was prepared from contributions by outstanding scientists and engineers in the radiation effects field, and
contains detailed design information for all types of electronic parts and circuits.
2. Dr. G. Keister and Boeing Company personnel, Data Book for Circuit Analysis
and Design-TREE, Vols. I and II (1964); obtain from Air Force Weapons Laboratory, Report #WLTDR-64-60. A valuable source of numerical data for the task

of designing radiation-hardened electronic systems. Values of empirical constants for use in the design equations given in this book may be found in Keister's data book, as well as the method by which the empirical constants were measured. The data book is also a valuable complement to D. C. Jones [1].

3. R. K. Thatcher, D. J. Hamman, W. E. Chapin, C. L. Hanks, and E. N. Wyler, "The Effect of Nuclear Radiation on Electronic Components," Radiation Effects Information Center Report No. 36, Battelle Memorial Institute (Columbus, Ohio), October 1964. Should be obtained on a mailing list basis by anyone doing radiation-hardening electronic design. It presents numerical degradation data for all types of electronic parts. It is continually updated as new data become available to REIC from industry, and is supplemented by a yearly report entitled "Radiation Effects State-of-the-Art." An example of this yearly report is REIC Report No. 34, June 1964.

4. F. W. Poblenz, "Analysis of Transistor Failure in a Nuclear Environment," IEEE, NS-10, No. 1, 74-79 (January 1963).

5. W. H. Sullivan and J. L. Wirth, "Methods for Measuring and Characterizing Transistor and Diode Large Signal Parameters for Use in Automatic Circuit Analysis Programs," IEEE, Radiation Effects Meeting, July 1965 Sandia Report SC-R-65-941.

6. P. A. Trimmer, "Transient Radiation Effects in Sample Triode Amplifiers," paper presented at IEEE Radiation Effects Meeting, April 1963.

BIBLIOGRAPHY

H. L. Olesen, "Nuclear Radiation Effects Design Guide for Electrical and Electronic Engineers," General Electric Technical Information Series No. 64SD243 (April 1964).

H. L. Olesen, "Designing Against Space Radiation," Electronics, pp. 61-71, December 28, 1964 and pp. 70-76, January 11, 1965.

GE Tube Department, "Research on Radiation Resistant High Temperature Thermionic Circuitry," Air Force Avionics Laboratory Technical Documentary Report #AL TDR 64-187 (August 1964).

R. F. Shea (ed.), IEEE, NS-11, No. 5 (November 1964).

Index

A

Absorbed dose, definition, 19
Absorbed-dose rate, definition, 19
Accelerators, linear, 151
Activation data, table of, 170, 171
Active device selection for radiation-hardened design, 66, 67, 74, 196-199
 semiconductor devices, 66, 67, 74, 196
 vacuum tubes, 196-199
Airborne scanner
 construction, 173
 operation, 173
Alpha cutoff frequency f_{aco}, 63
Alpha particles
 definition, 17
 range, 106
Analog circuit design, 211-215
 displacement effects hardening, 211, 212
 transient ionization effects hardening, 214, 215
Analysis
 electronic system, 191, 192
 failure probability, 189, 201-205
 material by neutron activation, 172
 radiation types, 27-32
 transient ionization effects, 197, 198, 205, 206
Analysis, circuit, 192, 193
 mathematical models for, 193, 205
 Charge-Control, 193
 Ebers—Moll, 193, 205
 Linvill, 193, 205

Annealing, of displacement damage, 41, 60
Antenna, Mariner II, 178, 179
Area monitoring, radiation exposure protection, 115
Atom
 interstitial, 55, 58
 knock-on, 57
 primary recoil, 57
Auroral radiation
 aurora borealis, 5
 injection mechanism for, 5
 magnetic latitude of, 5
 protons, 6
 yearly ionization dose, 5

B

barn, definition, 19
Beta-degradation equation, 63
Beta particles or rays
 definition, 17
 range, 110
Breakdown voltage of transistors
 conductivity sensitivity, 60
 impurity sensitivity, 60

C

C_{ST} (Stoermer unit), 124, 125
Cadmium cutoff, definition, 22
Cadmium ratio, definition, 22
Carrier lifetime τ, 49, 59, 53, 58
 definition, 49, 50
 minority, 50, 58

Cascade, neutron-produced displacement, 56
Chemical radiation effects, 42, 43
  effects on molecular configurations, 42, 43
  polyethylene, 43
  step-by-step process, 42
  Teflon, 43
Circuit analysis, 192, 193
  mathematical models for, 193
    Charge-Control, 193
    Ebers—Moll, 193, 205
    Linvill, 193, 205
Circuit design, 193, 194
  analog, 211-215
  circumvention, as a design technique, 192
  common mode rejection, 213
  design considerations, 98-100
  design techniques, 193
  digital, 207-211
  packaging techniques, 193
  selection of parts, 193
CIRCUITRON, 197
Cobalt glass, X-ray dosimeter, 154
Collective radiation exposure during man's lifetime, 108
Collector junction area of transistor, method for calculating in planar and mesa transistors, 82, 83
Common mode rejection, as a design technique, 213
Communication system, Mariner II, 177-179
Composite defect, definition, 58
Compton effect, definition, 17
Compton electron, 17
Conditions influencing experiment design, 137
Conduction band, 46, 47
Conductivity σ, in semiconductor, 48
Constant, damage (K_r), 58, 63, 66
Controlling the experiment, 137
Conversion table for ionizing radiation units, 26
Cosmic dust detector, Mariner II, 178, 180
Cosmic radiation
  constituents and energy, 1
  ionization dose attributable to, 1

Cosmic ray spectrometer
  construction, 163, 164
  operation, 163, 164
Cosmic rays, viii
  galactic, 158, 159
  solar, 159
Cosmology, 158-161
Crystal structure, 46-50
  formation, 48-50
Crystal surface, influence on device behavior, 48, 49

D

Damage constant K_r , 58, 63, 66
  doping sensitivity, 66
  surface sensitivity, 66
25% Damage dose, definition, 19
Degradation characteristics of transistors, 61-67
Degradation curve (transistor current gain), 61, 62
Density, solar wind, 6, 7
Depletion-region-controlled devices, 67-73
Design
  discussion, 195-206
  examples, 206-223
  handbook references, 223, 224
  procedure, 190-195
    circuit analysis, 192, 193
    circuit design, 193, 194
    electronic system, 191, 192
    nuclear radiation environment, 190, 191
    shielding, 191
    steps for, 190
    testing, 194, 195
$1/v$ Detector, 22
Diffusion, 51-53
Diffusion constant D, 51, 52
Diffusion length L
  method of calculation, 53
  related to transistor operation, 61
Diffusion length damage constants, solar cells, 74, 75
Digital circuit design, 207-211
  against displacement effects, 207
Diodes
  back-biased, 78, 79

transient ionization effects in, 78, 79

current flow in, 72

displacement effects in
rectifier, 73
switching, 74
Zener, 74

displacement effects on saturation current I_s and IR drop, 72

transient ionization effects in, 77

Dislocations, definition, 58

Displacement cascade, neutron-produced, 56

Displacement damage, effect on conductivity, 49

Displacement damage annealing, 41

Displacement damage by moving atoms, 40

Displacement damage threshold energy E_d, 48, 49

Displacement effect insensitivity, as a design technique, 215

Displacement effects, 54-59
displacement-induced defects, 41, 42
electrical resistivity, 41
mechanical properties, 41, 42
semiconductor effects, 41
thermal properties, 42
energy threshold E_d for, 40
magnetic materials, 202
on various semiconductors, 59-76
process for
displaced atom, 39
lattice defect movement, 39
nuclear particle, 39
rectifier diodes, 73
silicon-controlled rectifiers, 74
solar cells, 74
summary of all devices, 198
transistors, 61-72
tunnel diode, 74, 75
versus environment, 44, 45
versus the primary particle, 55-58
Zener diode, 74

Displacement mechanism, 57

Displacement radiation effects, 38-42
production by charged particles, 39
production by gamma radiation, 39
production by neutron radiation, 39, 40

Displacement spike, definition, 55

Dose, exposure,
rate, 19, 25
unit, 25

Dose rate contributions, 128

Dose rate monitors, 155

Doses, maximum permissible for radiation protection, 113, 114

Dosimetry, 152-155
gamma and X-rays, 154, 155
neutrons, 152-154

Dosimetry errors, neutrons, 153

Drift, 52
drift velocity $\bar{v}_d$, 52

Dual transistors, as a design technique, 207

E

E_d, displacement damage threshold energy, 48, 49

E_K, kinetic energy, 56

Ebers—Moll mathematical model, 193, 205

Electrical storage time t_s, 82, 83
method for measurement of, 82, 83

Electromagnetic pulse effect, 91, 92

Electromagnetic radiation
from the sun, 12
distribution of the energy, 12
energy, 12
units for, 23

Electron de-excitation, 38
process for, 38

Electron excitation, 37
electron—ion pair production, 37
process for, 37
radiation energy expended, 37

Electron spectrometer
construction, 162, 163
operation, 162, 163

Electronic materials radiation sensitivity, 198

Electronic parts selection, 195, 196
active devices, 196-199
passive devices, 199-201

Electronic system analysis, selection of design approach, 191, 192

Electronic system design techniques, 189-224

Electrons
  displacement effect threshold, 27
  displacement per centimeter, 57
  graph, displacement/centimeter, 57
  in Van Allen belt, 5
    energies, 5
    interactions, 5
  interaction mechanism, 57
  maximum energy transfer, 55-57
Elements, the search for, 160
Energy absorption, 19
Energy-level diagrams
  insulator, 46
  intrinsic semiconductor, 46
  metal, 46
  p- and n-type semiconductor, 46
Energy of sublimation E_c, 48
Energy required for ionization in semiconductor, 77
Energy spectra, neutrons, 140, 141
Energy threshold value for displacement damage E_d, 40
Environment simulation
  after circuit design, 195
  prior to circuit design, 194, 195
Environmental requirements for nuclear instruments in space, 161
Environments
  description of, 1-16
  radiation, 1-33
Epicadmium, range of neutron energy,
  20
Epithermal, range of neutron energy,
  20
Equal energy—equal damage concept, 25
Equation of continuity, 78
ergs/g(C), 23, 24
  reference measurement method, 24,
  25
ergs/g (of material m_0), 19
eV/g, 23, 24
Excess carrier generation equation, 77
Excited electrons, effects of, 37, 38
  in gases, 38
  in insulators, 38
  in semiconductors, 38
  in transparent materials, 38
Experimental considerations, 136-141
Experimental design, 137
Experimental facilities, 131-136

Exposure dose, 19
Exposure-dose rate, 19, 25
Exposure-dose unit, 25

F

F centers, 38
  production of, 36
Facilities
  pulsed radiation, 142-150
  steady-state gamma, 132, 133, 136
  steady-state gamma—neutron, 135,
  136
  transient ionization, 150-152
Failure probability analysis, 201-205
  importance of, 189
Fast fields, neutron, method for re-
  porting, 21
Fast neutrons, energy range of, 20
Feedback, as a design technique, 211,
  212
Fermi level, 46, 47
Field-effect transistor
  insulated-gate type, 71
  MOS, 67, 70, 71
  operation, 67, 69-72
  unipolar, 67, 70
Flares, solar, 6
Flash X-ray facilities, 151, 152
Foils, neutron dosimetry, 152, 153
Forbidden gap, definition, 46
Forward current transfer ratio β, 61
Frenkel pair, 58
Functional threshold dose, definition,
  19

G

G-value, 24
Gamma and neutron heating, 92-96
Gamma and X-radiation, 106
Gamma and X-rays, dosimetry, 154,
  155
Gamma exposure, method of describ-
  ing, 24, 25
Gamma-induced charge, 96, 97
Gamma radiation output, from radio-.
  isotope generator, 12
Gamma radiation unit conversions, 24,
  25
Gamma ray shielding, 122
Gamma ray test facilities

pulsed, 148
steady-state gamma, 132, 133, 136
steady-state gamma–neutron, 135, 136
transient ionization, 150-152
Gamma rays
analysis of, 29
definition, 18
high-energy, 159
use as direction indicators in the universe, 159
low-energy, 159
photoelectric reaction, 29
Geiger–Müller tubes, Mariner II, 178, 179
Gene mutations, 112
Generation term g, 77, 78
Godiva II reactor, 145-147
Ground Test Reactor (GTR)
characteristics of, 139-141
neutron energy spectrum, 141

H
Hall measurement, 53, 54
Hall coefficient R, 54
method of calculation, 54
Hall mobility μ_H, 54
Heat deposition rate, 93
Hereditary effects
function of dose rate, 114
gene mutations, 114
High-energy gamma ray telescope, 166-169
background, 166
construction, 167
operation, 167
Hole, as a carrier, 48

I
I_{pp}, primary photocurrent, 78, 79, 81
I_{sp}, secondary photocurrent, 84
Impurities' influence on semiconductor characteristics, 60
Impurity atoms, 55
Impurity density, graded, 60
Individual monitoring, radiation protection, 114, 115
Instrumentation designs, nuclear, 161-177

Instrumentation system, Mariner II, 177-181
Intensity, definition, 19
Intermediate neutron energy range, 20
Interplanetary plasma (solar wind), 160
Interstitial atoms, 55, 58
Ionization by electrons, 28, 29
Ionization effects
by space radiation, 10, 11
definition, 77
surface, 87
transient, 36, 37, 45, 150-152, 197-201
Ionizing radiation units, table, 26
Irradiated materials, radiation output from, 2

K
K_r
definition, 58
typical values, 66
KEWB A and B pulsed reactors, 142, 143
Kinetic energy E_K 56
Knock-on atom, definition, 57
KUKLA reactor, 147

L
Lifetime τ, definition, 50
Linear absorption coefficient (LAC) or μ, 18
Linear accelerators, 151
Linear energy transfer versus RBE, 108, 109
Linvill mathematical model, 193, 205
Low-frequency response, as a design technique, 209

M
Magnetic materials, displacement and other effects, 201, 202
Magnetic shielding, 122-124
Magnetosphere
influence by solar wind, 8
magnetopause, 8, 9
shape and extent of, 3, 8
spacecraft flights, 9
spacecraft investigations into the extent of, 8, 9
Man, radiation effects on, 100-115

Mariner II, 177–181
  antenna, 178, 179
  communication system, 177–179
  cosmic dust detector, 178, 180
  experiment description, 180, 181
  experiment results, 180–183
  Geiger–Müller tubes, 178, 179
  instrumentation system, 177–181
  magnetometer, 178, 179
  solar wind detector, 178, 180
Material analysis, by neutron activation, 172
Material effects, summary of, 97–100
  capacitors, 97, 98
  gas tubes, 97
  microwave tubes, 97
  vacuum tubes, 97
Mathematical models
  Charge-Control, 193
  Ebers–Moll, 193, 205
  Linvill, 193, 205
Maximum energy transfer, neutrons, 57
Maximum levels, Van Allen belt, 4
Maximum permissible doses, 113, 114
MeV/cm^2, 23
MeV/cm^2-sec, 23
Microplasma, 90, 91
Million electron volts (MeV), 18
Minority carrier lifetime, 50
  degradation equation, 58
Mission in space, 177–187
Mobility μ, 52, 53
  in semiconductor, 48
MOS (metal oxide semiconductor) transistor, 67, 70, 71
  ionization effects in, 84, 85

N
Neutron dosimetry, 152–154
Neutron energy spectra, types of, 140
Neutron foils, types of, 153
Neutron Phoswich
  construction, 169
  operation, 169, 172
Neutrons
  average displacement number, 56
  definition, 17
  fast fields
    fission and resonance foils, 22
    method for reporting, 21
  interactions by, 31

maximum energy transfer, 57
  shielding against, 126
Noise insensitivity, as a design technique, 209
Nuclear and particle physics, 161
Nuclear explosion, energy released, 16
Nuclear instrumentation designs, 161–177
Nuclear instruments, for particle physics, 175–177
  semiconductor particle detectors, 176, 177
  spark chamber, 175, 176
Nuclear instruments, uses for, 157–161
Nuclear power reactors
  power output, 12–14
  radiation from, 14
  typical exposure for an electronic system, 14
Nuclear radiation environment, 190, 191
Nuclear "reactor-in-flight" system parameters (SNAP system), 13
Nuclear rocket, transient ionization effects from, 15
Nuclear space systems
  nuclear power reactors, 12–14
  nuclear propulsion rockets, 15
  radioisotope generators, 12
  nv and nv_0, definitions, 22, 23

O
Observatories in space, 181–187
  OGO, 181, 185, 186
  OSO, 181
Orbiting observatories, 181, 184
ORNL Super Godiva reactor, 149, 150

P
Packaging design, 206
Particle energy, relation to damage production, 58
Particle equilibrium, 20
Particle flux in Van Allen belt, 4, 5
  heart of inner zone, 5
  heart of outer zone, 5
Particle interaction parameters, 28
Passive devices, radiation effects on
  capacitors, 200, 201
  magnetic materials, 201
  resistors, 199, 200

Passive shielding, 119-122
Pentavalent impurities, definition, 47
Photon, definition, 17
Photon/cm^2, 23
Photon/cm^2-sec, 23
Photon flux, definition, 19
Physics, semiconductor, 46-48
Pneumatic controls and logic, 223
Point defect, definition, 58
Power consumption trade-off, 216, 217
Primary photocurrent I_{pp}
  calculation of in a transistor, 81
  calculation of in back- and forward-
    biased diodes, 78, 79
  definition, 78
Primary recoil, 57
Protection standards, human exposure,
    113-115
Protons
  definition, 17
  energies, 4, 6
  in Van Allen belt, 4, 6
  interaction mechanisms, 56
  interactions by, 30
  mean number of atoms displaced per
    centimeter, 56
  shielding against, 126
Pulsed radiation facilities, description
    of, 142-150

Q
Quantum tunneling device, permanent
    effects on, 74, 75

R
rad, definition, 18
Radiation, reduction of from a nuclear
    space system, 15
Radiation effects
  on man, 100-115
  on semiconductor devices, 43-91
    basic processes, 44, 45
    other, 91-97
    types, 35-43
      chemical, 35, 42, 43
      displacement, 35, 38-42
      transient, 35-38
Radiation emitted from radioactive ma-
    terials, 109, 110
Radiation environment, produced by
    nuclear burst
  environment versus altitude, 15, 16

nuclear output, 15
shock and blast output, 15
thermal output, 15
Radiation environments, 1-33
Radiation experimentation, 131-136
  radiation environment versus simu-
    lation facility, 131
Radiation exposure and biological
    effects, 105-108
Radiation fields versus energy ab-
    sorbed, 25, 26
Radiation heating, 92-96
  heat deposition rate, 93
  interactions, 93
Radiation levels
  in a spacecraft from two solar flare
    events, 129
  in a typical spacecraft, 127-129
Radiation output from
  accelerators, 2
  fission fragments, 2
  irradiated materials, 2
  natural radioactive materials, 2
  reactors, 2
  space, 2
Radiation protection standards, 113-115
Radiation shielding, 119-130
  active, 122-124
  passive, 119-122
Radiation sources
  comparative risks from, 108-110
  typical, 2
Radiation storage time t_{sR}
  definition, 83
  method of calculation, 83
Radiation terminology, 16-27
Radiation test, TIMM record amplifier,
    221, 223
Radiation types, analysis of
  electrons, 27
  gamma rays, 29
  neutrons, 31
  protons, 30
Radioisotope generators
  gamma radiation from, 12, 13
  power output, 12
Rate of energy absorption, 19
RBE (relative biological effectiveness),
    107
RBE values, 107

Reactors
  Godiva, 145-147, 149, 150
  GTR, 139-141
  KUKLA, 147
  SPRF, 147, 148
  steady-state, 135, 136
  summary of pulsed, 148, 150
  TRIGA, 144, 145
Recombination center, 50, 51
Recombinations in GaAs and InAs, 51
Record amplifier, TIMM, 215-223
Rectifier diodes, displacements effects
    in, 73
rem (roentgen equivalent man), defini-
    tion, 18, 107
rep (roentgen equivalent physical),
    definition, 19
Resonance, neutron cross section, 20
RIFT, reactor-in-flight, 13
Roentgen, definition, 18
Roentgen per hour, definition, 18

S

Scintillation detector
  construction, 166
  operation, 166
Secondary photocurrent I_{sp}
  definition, 84
  method of calculation, 84
Secondary radiation, produced in shield
    by primary radiation, 126
Selection of transistors, 67
  base transit time, 67, 68
  minimum beta selection, 67
  transit time curve slope, 67
Semiconductor, forbidden gap in, 47
Semiconductor operation, 45-54
  crystal structure, 48-50
  diffusion and drift, 51-53
  lifetime, 50
  recombination centers, 50, 51
  semiconductor physics, 46-48
  traps, 50, 51
Semiconductor particle detectors
  construction, 176
  operation, 176
Semiconductor physics, 46-48
Semi-Rad
  construction, 172
  operation, 155

Shielding, 191
  active, 122-124
    electrostatic, 122, 123
    magnetic, 123, 124
  against long-term space radiation,
      124-129
  passive, 119-122
    gamma radiation, 120-122
    neutrons, 119, 120
Short-circuit gain a, 61
Silicon-controlled rectifiers, displace-
    ment effects in, 74
Silver-activated phosphate glass, X-ray
    dosimeter, 155
Slope parameter b, definition, 203
Slot, in Van Allen belt, 3
Slow neutrons, energy range, 20
SNAP, 13
Solar cells, displacement effects in, 74
Solar flares
  frequency of occurrence, 6
  ionization dose from, 6
  proton flux, 6
  solar storms, 6
  time of arrival at the earth, 6
Solar wind
  constituents of,
  density and flux, 6, 7
  effects on comet tails, 7
  extent of solar wind influence, 7
  influence on magnetosphere, 6
  velocity, 7
Somatic effects, 104, 111, 112
  acute, 111
  delayed, 111
  summary of, 111, 112
Space-charge distributions, relaxation
    of, 38
Space environment, near and solar, 1-
    12
Space radiation ionization and displace-
    ments caused by
  cosmic rays, 11
  inner Van Allen belt, 10
  outer Van Allen belt, 10
  solar flares, 11
  steady solar output, 11
Spacecraft construction, Mariner II,
    177, 178

Spark chamber, 174-176
  construction, 174-176
  history, 175
  operation, 175
Spectrometers, 162-165
  cosmic ray spectrometer, 163, 164
  electron spectrometer, 162, 163
  X-ray spectrometer, 164, 165
Spectrum, of particle field, 59
  electrons, 59
  neutrons, 59
  protons, 59
Spectrum determination, 139
SPRF (Sandia Pulsed Reactor Facility),
    147, 149
Stars, proton-produced, 30
Storage time
  electrical, 82
  radiation, 83
Steady-state gamma facilities, 132, 133,
    136
Steady-state gamma-neutron facilities,
    135, 136
Stoermer unit C_{ST}, 124, 125
Summary of pulsed reactor facilities,
    148
Surface effects, 45, 85-91
  definition, 85
  dose-rate sensitivity, 85, 87
  dose sensitivity, 85, 87
  manifestation, 85
  process sensitivity, 90
  selection of good performers, 90
  versus environments, 45
Surface recombinations, 51
Switching diodes, displacement effects
    in, 74
System analysis, electronic, 191, 192
System weight trade-off, radiation-
    hardened electronic system, 216,
    217

T
Table of activation data, 170, 171
Teflon, radiation effects in, 43
Test monitoring, 138, 139
Thermal neutrons, energy range of, 20
Thermal spike, definition, 55
Threshold dose, definition, 19

Time-integrated dosage, definition, 19
Timers, 223
TIMM components, 216-218
TIMM record amplifier, description
    of, 215-223
  construction, 220
  operation, 218, 219
TIMM (Thermionic Integrated Micro
    Module), description of, 196
TLD, X-ray dosimeter, 155
Total dose ratio, 139
Transient analysis, 205, 206
Transient ionization effects
  de-excitation of electrons, 36, 37
  excitation of electrons, 36, 37
  in capacitors, 200, 201
  in resistors, 199, 200
  in vacuum tubes, 197, 198
  perturbations, 37
  testing facilities, 150-152
  versus environment, 45
Transistors
  beta-degradation nomographs, 63-65
  displacement effects on a, 61
  operation, 49, 50
  prediction of transient ionization
      effects based on electrical meas-
      urements, 80
  selection of good performance in a
      particle environment, 67
  transient ionization effects in, 79-84
    analysis models, 80
    transistor not saturated, 80, 81
    transistor saturated, 80, 81
Traps, in semiconductor crystal, 50,
    51
TRIGA reactors, 144, 145
  Mark I, 144
  Mark II, 144
  Mark F, 144, 145
  pulsed operation, 145
Trivalent impurities, 48
Tunnel diode, displacement effects in,
    74, 75

U
Use of high-energy gamma rays as
    direction indicators in the uni-
    verse, 159

V
Vacancy, radiation-induced, definition, 55
Vacuum tubes, 196-199
  filament, ceramic, 196, 197
  filament, glass, 196, 197
  heaterless, ceramic, 196
Valence band, definition, 46
Van Allen belt
  altitude, 4
  contour plot of, 3
  description of, 2-5
  electrons in, 5
  inner, 4
  July 9, 1962 nuclear burst, 4
  low particle flux slot, 3
  magnetic latitude, 4
  maximum levels, 4
  outer, 4

particles, mechanism of generation, 160
  protons in, 4
  slot, definition, 3

W
Weibull distribution function, 201, 203-205
Wigner threshold energy E_d, 55, 56

X
X-ray observatory design, 186, 187
X-ray spectrometer, 164, 165
  construction, 165
  operation, 165

Z
Zener diodes, displacement effects in, 74